民以食爲天
中国饮食器物设计作品集

Food is God

Design of Chinese Dining Utensils

【主编】王立端 段胜峰

Compiled by Wang Liduan Duan Shengfeng

重庆大学出版社

Design of Chinese Dining Utensils

前言

| 王立端

作为普通老百姓，每天开门需要面对七件事情——柴、米、油、盐、酱、醋、茶。所以，在中国自古以来就有"民以食为天"的说法。设计的宗旨是服务于生活，可见，设计与饮食文化的关系密不可分。例如：食材的存放与加工，饮食的烹制与享用，就餐的氛围与礼仪，赏茶的习俗与品位，但凡与吃喝有关的一切无不与器物相关，食因器而增色，茶因器而添香。千百年来，正是由于饮食器物的发展促进了食材加工、保存，乃至营销方式的发展进步，创造出了人们可以健康进食、愉悦餐饮的诸多条件与方式，从而使饮食文化不断得以丰富。这其中，设计从功能、方法、形式诸方面发挥了至关重要的作用。事实上，设计自始至终从未停止对饮食器物设计创新的探索。

这册作品集所刊录的是一批设计者针对饮食器具造物的个人作品，这些作品制作工艺丰富、材料种类多样、视觉表达新颖。透过这些作品我们可以感受到作品创作者们或致力于材料语言的拓展；或专注于如何将传统手工艺和民间生活智慧应用于当代生活的探寻；或尝试运用现代加工技术通过作品追求返璞归真的境界。也许它们只是作者们的阶段性实验作品，只是中国饮食器物大海中的沧海一粟，但透过这些作品整体所呈现出的气息，我们仍然可以了解到中国饮食器物设计在传承中的不断进取、发展；在发展中通过广泛涉取来自古今中外营养后的创新；在创新中紧贴百姓的居家所需、审美所需、精神所需的各种作为。

设计是一种文化形式，而文化交流是不同地域之间、不同民族之间、不同国度之间相互沟通了解的最好方式。将这些作品汇编成册，并随同作品实物一起在中国、在欧洲、在世界展出和交流，是希望能让更多的人加深对中国饮食文化的了解，加深对中国饮食器物文化的了解，也加深对中国当代设计在继承中如何发展的了解。

在中国国家艺术基金的资助下，这些作品继2015年在意大利米兰、威尼斯展出之后，将于2017年赴法国巴黎、比利时布鲁塞尔、土耳其伊斯坦布尔等地巡回展出。"丝路漫长·器邀四方"，通过作品的展出与学术交流，我们期待赋予观众另一种感悟和了解中国文化的形式通道，让世界上更多的人透过作品"见微知著"地加深对中国文化的喜爱和认同，增加中国人民与世界人民之间的了解和交流，实现"品味中国、牵手世界"的策展目的。

Preface

| Wang Liduan

Common people all have to work every day for seven necessities of life—firewood, rice, oil, salt, soy sauce, vinegar, and tea. Accordingly, there goes the old Chinese saying "Food is the first necessity of man". Design is made for life, and it follows that design is interrelated to culinary culture, such as the storage and the processing of food, cooking and dining, the dining atmosphere and table manner, as well as tea culture. All things about dining and drinking are related to tableware, suitable tableware and tea set accentuate food and tea. Thanks for the advance of tableware in the past thousands of years, the processing and storage of food and even the marketing methods have been progressing and have created many ways of healthy and enjoyable dining and drinking, enriching our culinary culture constantly. In the progress, design has played a part of great importance in terms of function, methods, style of tableware. As a matter of fact, design has never stopped the exploration of innovation in tableware design.

This album is a collection of some designers' works of tableware, which present rich craftsmanship, diverse materials, and visual novelty. In these works exhibit designers' commitment to express their ideas with various material languages, to explore the ways to utilize in contemporary life the wisdom from traditional handicraft and folk life, and to pursue the realm of simplicity and basics through modern processing technology on materials. Maybe, they are just designers' experimental works in the current stage and a drop of water in the sea of Chinese tableware culture. However, what expressed in these works is a momentum of the development of tableware design with the heritage of Chinese culture, the innovation upon the collective wisdom of all times and all over the world, the efforts for people's home life, artistic demand, and spiritual need.

Design is a form of culture, and cultural exchange provides the best way to link up different regions, different peoples, and different nations. The album of these designs along with the exhibitions of the objects in China, in Europe, and around the world is in hope of letting the world know more about Chinese culinary culture, Chinese tableware culture, and also the development of Chinese contemporary design in cultural heritage.

Funded by China National Art Fund, these works had their exhibitions in Milan and Venice, Italy in 2015 and will be on tour in Paris, France, Brussels, Belgium, and Istanbul, Turkey in 2017. "On the endless journey of Silk Road are various cuisines entertained." Through the exhibition and exchanges, we hope to provide our audience a different reflection on Chinese culture and a visual channel to know it, win through a glimpse of exhibits more popularity and recognition around the world for Chinese Culture, link up Chinese people and the rest of the world, and make the exhibition a platform to serve Chinese flavor and make friends with people around the world.

"丝路长·宴四方——中国饮食器物设计文化展"巡回展览路线图

The Road Map of the tour "Feast along the Silk Road—An Exhibition of Chinese Tableware Design and Culture"

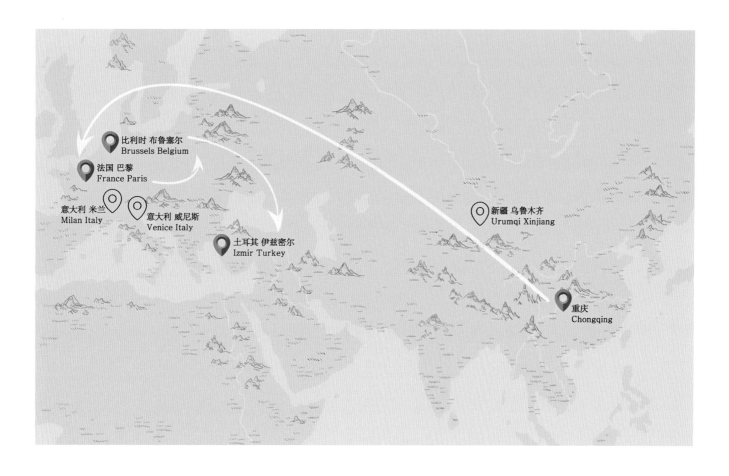

目录 Contents

P10~15
中国人的"味道"
——全球化时代的最后一道文化壁垒 /
Chinese Philosophy of "Flavor"
——The Last Cultural Castle in the Global Age
谢亚平 /Xie Yaping

P18~19
巢鸾共觅 /Nest and Phoenix
陈俊平 /Chen Junping
谭忠诚 /Tan Zhongcheng

P20~21
"八大碗" /
Eight Traditional Dishes
唐英 /Tang Ying
李洲洲 /Li Zhouzhou
李玥洋 /Li Yueyang
包钰 /Bao Yu
张振环 /Zhang Zhenhuan

P22~23
锔瓷造器之一 /Ju-China Ware Ⅰ
张建平 /Zhang Jianping
罗显怡 /Luo Xianyi

P24~26
锔瓷造器之二 /Ju-China Ware Ⅱ
张建平 /Zhang Jianping
罗显怡 /Luo Xianyi
工艺支持：王天昊 / Wang Tianhao

P27
陶与瓷 /Pottery and Ceramics
唐英 /Tang Ying
李洲洲 /Li Zhouzhou
李玥洋 /Li Yueyang
包钰 /Bao Yu
张振环 /Zhang Zhenhuan

P28
漆息·气息 /Paint Atmosphere
张国栋 /Zhang Guodong
张津亚 /Zhang Jinya

P29
水泥与木 /Cement and Wood
傅淑萍 /Fu Shuping
颜璐 /Yan Lu

P30~31
四季 /Four Seasons
邓莲 /Deng Lian
王立端 /Wang Liduan

P32~34
意陶二十四器 / "24" Pottery
尧波 /Yao Bo

目录 Contents | 05

P35
他山之石 /Stones From Other Hills
李瀚然 /Li Hanran
蒲柯宇 /Pu Keyu

P36~37
相逢 /Meeting
周亚蓬 /Zhou Yapeng
王立端 /Wang Liduan
吴菡晗 /Wu Hanhan

P38
再造·重生 /Regeneration
何艳 /He Yan
陈静 /Chen Jing

P39
疯狂蔬果 /Crazy Creatures
罗黛诗 /Luo Daishi

P40
青渺 /Celadon Charm
范易 /Fan Yi
贾倩 /Jia Qian
郭桂林 /Guo Guilin
李尧瑶 /Li Yaoyao
袁雁辉 /Yuan Yanhui

P41
石器 /Creative Stone Ware
罗黛诗 /Luo Daishi
雷霞 /Lei Xia

P42~44
新概念泡菜坛 /Creative Pickle Jar
杨曼羚 /Yang Manling
邹红媛 /Zou Hongyuan
周丽雯 /Zhou Liwen

P45
馄饨皮 /Wonton skin
夏扬扬 /Xia Yangyang

P46~47
山海杯 /Mountain Sea Cup
王立端 /Wang Liduan
白晓宇 /Bai Xiaoyu
范易 /Fan Yi
何源源 /He Yuanyuan
张瀚文 /Zhang Hanwen
陈梦秋 /Chen Mengqiu
廖桦 /Liao Hua
任宇 /Ren Yu
李瀚然 /Li Hanran

P48~49
草草成器 /Folk wisdom
王立端 /Wang Liduan
叶凌杉 /Ye Linshan
吴菡晗 /Wu Hanhan

P50
器节 /Qijie
何源源 /He Yuanyuan

P51
皱 /Cun
何源源 /He Yuanyuan
王立端 /Wang Liduan
黄欣 /Huang Xin

P52~53
晴雪、家园 /Sunny & Snow, Homeland
张国栋 /Zhang Guodong
张津亚 /Zhang Jinya

P54~55
砂器 /Grit Utensils
廖桦 /Liao Hua

P56~57
无题 /Untitled
邓莲 /Deng Lian

民以食为天 Food is God

P58
竹节·丝语 /
Bamboo Joints and Filaments
任宇 /Ren Yu
谢睿 /Xie Rui

P59
咏竹 /An Ode to Bamboos
任宇 /Ren Yu
谢睿 /Xie Rui

P60~61
旋切木艺 /Rotary-cutt Wood Art
吴时敏 /Wu Shimin

P62
玲珑面点 /Delicate Pastry
范易 /Fan Yi
贾倩 /Jia Qian
郭桂林 /Guo Guilin
李尧瑶 /Li Yaoyao
袁雁辉 /Yuan Yanhui

P63
时食餐具 /Seasonal Food Tableware
赵卫东 /Zhao Weidong
彭科星 /Peng Kexing
刘倩娇 /Liu Jiaoqian
马林明 /Ma Linming

P64~65
夏布食器系列 /
Grass Linen Tableware Series
白晓宇 /Bai Xiaoyu
陈石 /Chen Shi

P66~67
竹编焐库 /
Bamboo-woven Insulated Container
王晨雨 /Wang Chenyu

P68~69
四川小吃器皿 /
Sichuan Snacks Ware
曹宇嘉 /Cao Yujia
张婷 /Zhang Ting

P70~71
沙岩 /Sandstone
王立端 /Wang Liduan

P72
贝·壳 /Shells
李肖依 /Li Xiaoyi
袁星来 /Yuan Xinglai

P73
缱 /Tie
蒲柯宇 /Pu Keyu

P74
食盒（一） /Hamper Ⅰ
何欢 /He Huan
张建平 /Zhang Jianping
罗显怡 /Luo Xianyi

P75
食盒（二） /Hamper Ⅱ
何欢 /He Huan

P76
漆皿 /Lacquerware
陈静 /Chen Jing
白静 /Bai Jing
李双利 /Li Shuangli
伍姝梅 /Wu Shumei

目录 Contents | 07

P77
凿木为器 /
Rough Wood Lacquer Ware
王立端 /Wang Liduan
李其龙 /Li Qilong
糜思尧 /Mi Siyao
贺杰 /He Jie

P78
惜之，缮之 /Cherish It, Repair It
张国栋 /Zhang Guodong
张津亚 /Zhang Jinya

P79
粒粒皆辛苦 /
Every Grain Comes From Hard Working
齐敏达 /Qi Minda

P80
山影 /Mountain Shadows
梁云云 /Liang Yunyun

P81
莫兰迪的漆物 /
Lacquerware of Morandi
刘利 /Liu Li
张媛媛 /Zhang Yuanyuan

P82
暖暖 /Warmth
廖桦 /Liao Hua

P83
安陶四合碗 /
Anfu Pottery Quadrangle Bowls
廖桦 /Liao Hua

P84
灰釉器皿 /Ash Glaze Utensils
张茜 /Zhang Qian

P85
食语 /Voice of Food
彭威 /Peng Wei
柳芳 /Liu Fang

P86
虚极与静笃 /
Detached and Unselfish in the Space
李颖 / Li Yin

P86
方 /Square
李其龙 /Li Qilong

P87
柱 /Columns
于瑛豪 /Yu Yinghao

P87
溯果 /Origin of Fruits
白玥 /Bai Yue
毛安澜 /Mao Anlan

P88
星石 /Asteriated Stones
温美婷 /Wen Meiting

P89
垣 /Walls
冯宇森 /Feng Yusen

P90
抹布 /Cloth
蒲柯宇 /Pu Keyu
李瀚然 /Li Hanran

P91
紫金茶具 /Red Golden Tea Set
陈学渊 /Chen Xueyuan

P92
合食 /Shared Meal
包钰 /Bao Yu
李洲洲 /Li Zhouzhou
刘檬 /Liu Meng
张振环 /Zhang Zhenhuan

P93
肌肤下的火焰 /
Flame under the Skin
尧波 /Yao Bo

P94~95
胎 /Base
文森(法国) /Vincent Cazeneuve

P96~97
清朴·提盒 /
Simplicity Carrier with a Handle
张瀚文 /Zhang Hanwen

P98
五谷丰登 /A Bumper Harvest
唐愚程 /Tang Yucheng

P99
褶皱盘 /Wrinkle Plate
王远秋 /Wang Yuanqiu

P100~101
阖家—锦绣食 /
The Whole Family – Numerous Delicacies
张一潘 /Zhang Yifan
李玥 /Li Yue
牟彦宣 /Mou Yanxuan
廖雯霏 /Liao Wenfei

P102~103
净桌 /Clean Table
吴菡晗 /Wu Hanhan

P104~105
晒桌 /Drying Table
赵宇 /Zhao Yu
糜思尧 /Mi Siyao

P106~107
草木缘 /Affection of Straw and Wood
敖进 /Ao Jin
杨承颖 /Yang Chengyin
刘潇 /Liu Xiao

P108~109
竹八仙 /Bamboo Square Table
张海涛 /Zhang Haitao

P110~111
新中式实木八仙桌之一 /
Neo-Chinese Style Solid Wood Square Table I
谢垚 /Xie Yao

P112
新中式实木八仙桌之二 /
Neo-Chinese Style Solid Wood Square Table II
谢垚 /Xie Yao

P113
交互石代 /Interactive Stone Age
吴菡晗 /Wu Hanhan
田棱锐 /Tian Lingrui

P114~115
高山流水 /Mountain & Stream
赵卫东 /Zhao Weidong
彭科星 /Peng Kexing
罗昊 /Luo Hao
马林明 /Ma Linming

P116~117
糙木茶桌 /Rough Wood Tea Table
王立端 /Wang Liduan
李其龙 /Li Qilong
糜思尧 /Mi Siyao

P118~119
舞 /Dance
张海东 /Zhang Haidong

P120~121
族 /Nationality
潘宏甲 /Pan Hongjia

P122
静直案儿 /Quiet and Straight Case
谢垚 /Xie Yao
张庆 /Zhang Qing

P123
沿 /Edge
李其龙 /Li Qilong

P124
组合展架 /Combination Display Rack
曹悦 /Cao Yue

P125
芸薹未了 /Gone Winter Rapes
汪梅 /Wang Mei
冯唐菡 /Feng Tanghan
曾贤思 /Zeng Xiansi
周璐丹 /Zhou Ludan
王鸿 /Wang Hong

P126~129
文化传承的手工艺教学创新实践 / Cultural Heritage Through the Innovation Practice in Handicraft Education
韦芳 /Wei Fang

中国人的"味道"——全球化时代的最后一道文化壁垒

| 谢亚平

　　味觉,是一种最顽固的文化基因,被镶嵌在每个人的DNA中。从儿时的记忆开始,到长大后归家,关于"味道"的谈论,成为一种集体共识,在中国人的文学艺术作品中常被提及和渲染。丰子恺在《给我们的孩子》里讲了儿时不能忘却的事,就是与父亲中秋赏月,而赏月之乐的中心,在于吃蟹。他记录了父亲的观点:吃蟹是风雅的事情,吃法也要内行才懂得。"先折蟹脚,后开蟹斗……脚上的拳头(即关节)里的肉怎样可以吃干净,脐里的肉怎样可以剔出……脚爪可以当作剔肉的针……蟹螯上的骨头可以拼成一只很好看的蝴蝶"。最初级的味道源于蟹肉对感官的刺激,是嗅觉、味觉、视觉的统一协作产生的一种体会;更高一级的味道来源于将蟹理解为风雅之物,植根于感官,但又进行了艺术升华。中国文学家常在其作品里表达这种升华。

　　当然,与之同时被重视的还有品味食物的方式。这种关于"吃"的方式,是一种最显性的文化样式,从食物的种植方式、处理方式、搭配原理、盛放方式到馈赠礼仪等,共同自觉地构筑成每个族群在全球化时代的最后一道文化壁垒。我们欣喜地品尝全球的美食,使用不同国家的餐具,但是,你的鼻子、你的胃、你的手,会不断地提醒你,你是谁?什么才是你最熟悉的味道?

一、文化圈层与饮食共生

　　中国的饮食文化历来丰富多样,这得益于中国幅员辽阔。复杂多样的地理环境和不断交融的历史文化,从热带到寒带,从海滨到高山,使这里物产丰富、食料广泛。

　　中国饮食研究专家赵荣光先生用12个饮食文化圈来表示饮食的地域性差别,这些饮食文化圈与中国历史文化圈基本重叠。他认为中国的饮食总体追求味道谐调和中,但不同地方的人口味却千差万别。交错的地貌和气候使得中国饮食文化里包含了明显的地域性差别,如西南以辣去湿、北方多食咸肉以御风寒、海疆岛屿则多食咸鲜海产、缺盐地区则以酸辣中和碱食。从北到南,口味由咸转淡;从西到东,口味由辣转甜;从陆到海,味道由重转轻。各饮食文化圈之间既相对独立,又相互渗透影响。

中国饮食文化追求个体生命与自然、社会、人际之间的和谐与均衡，这与中国文化崇尚"天人合一"的观念是一脉相承的。所吃食物首先要讲"天时"，与季节相适，挑选时令食物；还要讲"地气"，强调本地食物的选用。另外，还要重视饮食结构（谷、果、畜、菜）的和谐搭配；同时，中国人认为医食同源，养生保健，五味（酸、甜、苦、辣、咸）调和五脏，五脏和平则精神健爽。地域的饮食口味无论怎样变化，都隐含着调和五脏、健康养生的本要，隐含着中国思想文化的影子。

二、"美食"与"美器"相宜

丰富的"美食"还需与"美器"相配。中国饮食餐具的发展经历了原始陶器阶段、青铜器阶段、漆器阶段，发展至瓷器时代达到鼎盛。

中国古人甚至认为"美食不如美器"。饮食器具是生活器用最重要的物质载体。中国古人不仅关注美器，还强调使用的规范，"煎炒宜盘，汤羹宜碗，参错其间，方觉生色。"这种"生色"是器用功能与精美食物的相互衬托产生的审美大餐。清代文人张英在《饭有十二说》一文中提到，食器的选用需要让使用者有"自适之趣"。他说："器以瓷为宜，但取精洁，毋尚细巧。瓷太佳，则脆薄易于伤损，心反为其所役，而无自适之趣矣。予但取其中者。"他讲到食器的挑选以瓷为宜，选用精洁的瓷器，如果太薄，容易损伤，让使用者使用起来小心翼翼，有所障碍。这种审美向度体现了中国器物以"宜"为核心，精练而适宜，简约而另出心裁的工艺法则和价值标准。"宜"是一种理性与感性平衡的产物，是实用价值和审美价值之间的平衡，是中国传统艺术精神的产物。

《长物志》序中沈春泽亦强调"精而便，简而裁"。"简"不仅是一种文人士大夫崇尚的理想品格，也是工艺美学的重要内容，追求的是质的纯朴、精而简的风格。所谓以"宜简不宜繁"作为设计的尺度，既是经济的，又是审美的。"宜"的着眼点是人的实际可行的生活内容和尺度。这种造物之简，主要表现为一种审美和文化精神上的"简雅"追求。从实用的经济层面出发，倾达于本真之美。

中国广义的食器包括了盛餐具、炊具、储藏具等。炊具中蕴含的智慧也体现了中国文化追求"和"的特质。比如云南地区有种特殊的炊具——"汽锅"，这种炊具利用液化放热的原理设计而成。沸腾后的水汽在锅所构筑的感性空间中流动，成为一种独特的烹饪方式，锁住了浓烈的香味。同样，中国的烹饪讲究"锅气"，这是中式炒菜的精髓所在，充分利用锅热后再放油，油热后再放原料，大火快速爆炒等步骤，利用热气快速激发出食材的香味。

三、百姓日常与天道

中国人对吃的讲究已经深入骨髓，政治家管子曾说"王者以民为天，民以食为天"。"食"不仅仅是吃饱饭的问题，也隐含着政治与政权，与天道、哲学一样重要。因为在中国文化里，天道是蕴含在百姓每天都接触的平常事中。正如中国明代哲学家王艮提出的哲学命题，"百姓日用即道"。所以，"味道"一词在《辞海》就被解释为体味道的哲学。《易经》说"一阴一阳之为道"。研究者李苒认为，在中国饮食文化中的"味"包括鲜香两个方面，是阴阳之道的绝妙的形象诠释。

中国的饮食器物和进食家具中包含着中国人的自然观、物用观和人际观。比如明式八仙桌，强调一种和谐的"人际关系"，以拉近并均化进食空间距离的正方形或正圆形来规制。这种结构适应着，同时也规定着中国人"共餐"的进食方式，在"共餐"进食方式中启发、培养和造就以"和谐"为尚的人格品行。这种对群体关系的规制恰恰是通过吃饭这个行为所保留的一种独特的中国文化景观，成为全球化时代里最根深蒂固的文化信息。

Chinese Philosophy of "Flavor" — The Last Cultural Castle in the Global Age

| Xie Yaping

Taste, a most inherent cultural gene, is born with everyone. From the earliest memory of taste to the haunting memory of the flavor of home meal throughout one's life, everyone talks with joy about "flavor", a thing often mentioned and rendered in Chinese literature and arts. FENG Zikai in his book To Our Children relates the unforgettable thing of his childhood, enjoying the moon with his father, the best part of it being eating crab. He records in the book what his father told him about eating crab: "eating crab itself is artistic, and only a gourmet knows how : Tear down crab's legs, and then open crab bucket… the way to eat up every slice of the meat in the fist (or joint)of a leg, the way to pick out the meat in the belly…crab claw is just the pick for crab meat…the bones in the crab pincer can be pieced up into a beautiful butterfly. "When the crab meat causes one the sensory stimuli, an experience integrate of smell, taste, and sight, the senses form the base flavor of crab meat. When crab itself is regarded as an artistic thing, such a sublime makes the flavor of crab meat go beyond physical experience, a sublime often read in Chinese literature.

When flavor enjoys people's attention, the ways to enjoy food also catch their attention. From planting food, processing food, matching food, serving food to food gift etiquette, the ways to "eat" is an explicit cultural pattern, all the ways naturally forming a cultural barrier of a people in this global age. We enjoy global food and use table wares of other nations; however, our nose, our stomach, and our hands are reminding us who we are and what our most familiar flavor is.

Ⅰ. The Symbiosis of Culture and Cuisine

The vast land, the complicated and diverse geographical conditions(from tropic zone to frigid zone and from coastal region to high mountains), and the integration of other cultures in history, all breed a nation with abundance of natural resources and foods.

ZHAO Rongguang, an expert in Chinese cuisine study, differentiates regional foods with 12 spheres of food culture, the spheres mostly overlapping with the spheres of historical culture. According him, Chinese food features moderate flavor on the whole but diverse flavors regionally. The diversity of regional cuisine mirrors the diversity of land forms and climates throughout the nation. For instance, south-west cuisine features spicy flavor to drive off humidity, northern cuisine salty meat

to withstand bitter cold, coastal regions and islands seafood, salt deficient region sour and spicy food to compromise alkaline diet; salty gets less and less from north to south, and spicy turns into sweet from west to east; the heavy dressing becomes light dressing from mainland to island. In one word, every cuisine sphere is independent as well as inter-active with one another.

Chinese culinary culture seeks between individuals and nature, society, and interpersonal relation the harmony and balance, both of which are the extension of Taoism "man is an integral part of nature". Therefore, cooking is an art agreeable with seasons and places, in-season vegetables and local food being the best choice, the right proportion of various foods (grains, fruits, livestock, and vegetables) emphasized. Besides, Chinese deem that food has curing power and maintains health, that five flavors (sour, sweet, bitter, hot, and salty) agree respectively with five inner organs, the health of which makes one healthy and energetic from the inside out. No matter how different a regional food tastes, it sticks to the principle of health care, expressing implicitly Chinese culture.

II. Delicious Food with Delicate Tableware

Various delicious foods go with delicate tableware. The making of Chinese tableware evolved in phases of primitive pottery ware, bronze ware, lacquer ware, and lastly porcelain ware, the peak of utensil making.

Ancient Chinese even valued delicate tableware higher than delicious food itself, the tableware being the most important material carrier among all the other utensils in daily life. People not only care the appearance of the tableware but also emphasize the rules of using them, "pan-fried and stirred-fried food going with plates, thin soup and thick soup with bowls, and various dinner wares on one table making a feast not only to stomach but to the eyes of the beholder". "Feast to the eyes of the beholder" explains well that the perfect match between the tableware and the delicacy make an artistic sense of a feast. ZHANG Ying, a man of letters in Qing Dynasty, once mentioned in his Essay on the Twelve Appropriateness in Dining the choice of tableware is supposed to be suitable to the diner himself. He argues that "porcelain wares are good for dining, fine and clean ware being good enough, in case that refined and delicate wares deter the diner from dining for their fragility. I prefer the one suitable for me." In other words, the porcelain makes the best choice for tableware, a piece of fine and clean porcelain ware being preferable to the refined and delicate one because the latter is fragile and therefore make the diner too cautious to enjoy himself with the food. His criteria of choosing tableware exhibits the Chinese rule and norm of making tableware, all concentrated on "appropriateness", fine and suitable, simple but original. "Appropriateness" as a result of balance between reason and emotion and between pragmatic value and aesthetic value, is a product of Chinese traditional artistic sense.

SHEN Chunze also highlights in his foreword of Treatise on Superfluous Things the idea of being "fine and convenient, simple and suitable". "Simplicity" is not only an ideal personality upheld among scholar-officials but also an important matter in craft aesthetics, a pursuit of good quality and fine-simple style. The design principle of "simplicity rather than complexity" is both economic and aesthetic, focusing on the routine of daily life, expressed in a pursuit of

"simplicity and grace" in aesthetics and spirit, starting with pragmatism and ending up in beauty of simplicity.

Chinese tableware, in broad sense, range from serving wares, cooking wares to storing wares, all of which reveal wisdom and "harmony", an attribute of Chinese culture. For instance, "steam pot", a special cooker in Yunnan province, is designed based on vaporization theory, the vapor flowing in the concealed space and therefore sealing the aroma of food in the pot. And "stir fire", another term in Chinese cuisine, is the essence of Chinese stir-fried dishes, a cook heating the wok with high heat at first, then pouring into it oil, afterwards pouring into it ingredients and stir-frying them quickly with high heat to accentuate the aroma of ingredients.

Ⅲ. Daily Life and Natural Law

The delicate culinary has been an integral part of life of Chinese people. GUAN Zi, an ancient Chinese politician and philosopher, says, "His People matter to the king, and food matters to his people." According to him, "eating" is not only an issue of stuffing one's stomach but an issue of politics and powers, of the same importance with natural law because it is an integral part of daily life in Chinese culture. WANG Gen, a philosopher of Ming Dynasty, also proposes that "People's daily life is Tao(natural law)." Consequently, the word "flavor" is defined in Cihai, a Chinese dictionary, as a philosophy to experience flavor, and, in The Book of Changes, as "a combination of Yin and Yang". LI Ran, a researcher of Taoism, claims that the "flavor" in Chinese culinary culture is inclusive of taste and aroma, a greatly vivid interpretation of Yin and Yang.

The culinary wares and furniture also embody Chinese philosophy of nature, use of utensils, and interpersonal relations. For instance, Eight Immortals Table, a square or a round table of Ming Dynasty, shared by eight diners at the same time, emphasizes a harmonious interpersonal relation, a piece of furniture made to shorten and equalize the distance between the eight diners. Such an attribute adapts to and meanwhile determines the Chinese way of dining, or "share meal", which in turn inspires, cultivates, and shapes a Chinese characteristic of cherishing harmony with others. Such a traditional characteristic as a special Chinese culture is reserved through dining behavior, remaining a deeply-rooted Chinese culture standing erect in this global age.

《巢鸾共觅》/ 概念 "八大碗" 餐具系列
作者：陈俊平　谭忠诚
材质：陶瓷

Nest and Phoenix/ Conceptual "Eight Traditional Dishes" Tableware Series
Designers: Chen Junping, Tan Zhongcheng
Material: Ceramic

　　《巢鸾共觅》八大碗餐具设计，造型采用官帽、钱币、太阳鸟纹、人面纹、绳纹解构重组而成，展现"天美禹德,使百鸟还为民田"的农耕文化。形式上总体运用半球体、方圆结合，设计中檐口上扬，四足运用抽象的人面阴刻纹饰加以点饰，器型的曲面运用自然的绳纹与光面形成有机整体。配以独创的天然红丝线面釉，无铅无毒，耐酸耐磨，意在让人们在"年、节、庆典、迎、送、嫁娶"等八大碗宴请中远离污染，走进绿色、健康的优质生活。

The design of "Nest and Phoenix" "Eight Traditional Dishes" tableware is recombined by decomposing government official hats, coins, sunbird, face pattern and strand pattern, showing the agricultural culture of "the god makes birds to help build folk land". In the aspect of form, the tableware generally adopts combination of hemispheroid, square and round. The rim is rising, and four feet use face pattern for ornament. The curved surface of the tableware uses natural strand pattern, which forms organic whole with the shiny side. The unique natural red silk thread cover glaze is also applied, which is lead-free and non-toxic, and acid-resisting and wear-resisting. It means that people can be far away from pollution and can have green and health high quality life in fetes with "Eight Traditional Dishes" in events of "Spring Festival, Holidays, Ceremonies, Welcome Parties, Farewell and Marriage".

《"八大碗"之高脚器皿》/ 概念"八大碗"餐具系列
作者：唐英　李洲洲　李玥洋　包钰　张振环
材质：木、陶

Stemmed Ware for "Eight Traditional Dishes" / Conceptual "Eight Traditional Dishes" Tableware Series
Designers: Tang Ying, Li Zhouzhou, Li Yueyang, Bao Yu, Zhang Zhenhuan
Material: Wood, Pottery

作品灵感来源于汉代传统器型特征。依据中国传统"天圆地方"之理念，以及"民以食为天"的传统说法，采用圆口方底的器型。用拉坯成型和模具印制两者相结合的方法，用红陶泥制成，以白陶泥浆饰面，塑造出了具有传统文化韵味而又朴实大方的器皿。

The design idea of the work is from characteristics of traditional ware in Han Dynasty. In accordance with traditional Chinese idea of "round heaven and square earth", as well as the old saying of "food is the paramount necessity of the people", the tableware adopts shape of round mouth and square base. The tableware is made by red pottery combining the methods of throwing modeling and mould printing. The white pottery slurry is used for facing. The tableware with traditional culture and simplicity and generosity is created.

《锔瓷造器之一》/ 碎瓷重生器皿系列
作者：张建平　罗显怡
工艺支持：王天昊
材质：陶瓷、大漆、金箔、铜

Ju-China Ware I / Crackle Ware Series
Designers: Zhang Jianping, Luo Xianyi
Craft Support: Wang Tianhao
Materials: Ceramic, Lacquer, Gold foil, Copper

锔瓷是中华民族特有的一门修复陶瓷器皿的传统手工艺。运用这一传统技艺，不是为了修复，而是将不可再生降解的碎瓷片，经打磨拼接、锔合的同时进行器型再造，赋予碎瓷片新的生命和审美意义，让其具有独具一格的观赏性和艺术价值。

Ju-China is a unique traditional handicraft of the Chinese name to repair ceramic wares. This traditional handicraft is not applied to repair here, but to remold the wares by non-renewable and non-degradable via polishing, splicing and piecing so as to give new life and aesthetic significance to the porcelain pieces and give them unique ornamental value and artistic value.

民以食为天 Food is God | 24

《锔瓷造器之二》/ 碎瓷重生器皿系列
作者：张建平　罗显怡
工艺支持：王天昊
材质：陶瓷、大漆、金箔、铜

Ju-China Ware II / Crackle Ware Series
Designers: Zhang Jianping, Luo Xianyi
Craft Support: Wang Tianhao
Materials: Ceramic, Lacquer, Gold foil, Copper

中国饮食器物设计作品集　Design of Chinese Dining Utensils | 25

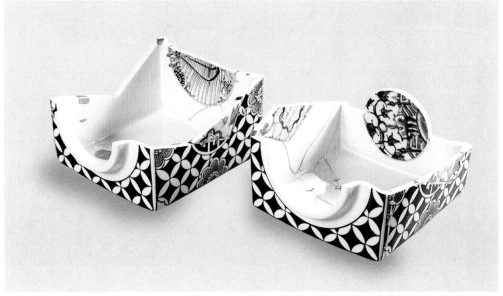

《锔瓷造器》/ 碎瓷重生器皿系列
作者：张建平　罗显怡
工艺支持：王天昊
材质：陶瓷、大漆、金箔、铜

Ju-China Ware / Crackle Ware Series
Designers: Zhang Jianping, Luo Xianyi
Craft Support: Wang Tianhao
Materials: Ceramic, Lacquer, Gold foil, Copper

作品灵感来源于地表岩石的肌理效果，采用拉坯成型的手法，选用白陶与红陶泥浆的结合，多次涂抹，形成自然开裂的装饰效果。釉色上选择丝光汶窑白，釉的温润素雅，与泥浆开裂后的粗糙感创造出独特的陶器魅力。

Inspired by the texture of surface rock, the designer, by using molding casting techniques, combines the multiple painting of red pottery clay mud with white pottery clay mud, reflects a naturally crack decorative effect, while choosing mercerizing Ru kiln as glazing color, the warm and elegant glaze and roughness of cracking clay mud create an unique charming of pottery.

《陶与瓷》/ 概念"八大碗"餐具系列
作者：唐英　李洲洲　李玥洋　包钰　张振环
材质：陶、瓷

"Pottery and Ceramics" / Conceptual "Eight Traditional Dishes" Tableware Series
Designers: Tang Ying, Li Zhouzhou, Li Yueyang, Bao Yu, Zhang Zhenhuan
Materials: Pottery, Ceramic

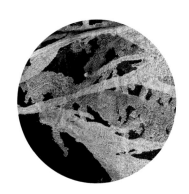

　　创作中偶然把漆灰和稀释剂撒到刚髹涂完的漆面上,出现了奇特的起皱现象。作者利用这一现象将调好的漆色泼在漆面上,让它随器型和起皱的肌理走向自由流动;漆与漆的交融,在被火灼烧后的漆面上随温度的骤变而自然定格;以漆灰为墨,借助变幻莫测的火势和漆性去绘制痕迹,顺应尘埃自然飘落其上;用笔蘸稀释剂对漆皱、漆液、漆灰产生不同的变化;擦拭、泼撒、熏灼,在失控与控制之间迂回创作,只为捕捉漆给予的那无可复制的瞬间气息。

During the creation process, the gray paint and the diluent happen to spray onto the painted surface, causing magical wrinkles. The phenomenon inspires us to pour the mixed color onto the paint surface so that the paint can flow on freely according to the texture of the vessel and the wrinkles. The mixed paint is fixed on the fired painted surface along with temperature changes. The gray ink is used as the ink to paint the traces according to the unpredictable fire behaviors and paint characteristics. The gray ink falls onto the painted surface naturally like the dust. The brush is dipped with the diluent to cause different changes of paint wrinkles, liquid paint and gray paint. By wiping, spraying and smoking, we move from being in control to being out of control so as to capture the irreproducible instant atmosphere of the paint.

《漆息·气息》
作者:张国栋 张津亚
材质:纸、大漆、瓦灰

Paint Atmosphere
Designers: Zhang Guodong , Zhang Jinya
Materials: Paper, Lacquer, Tile paint

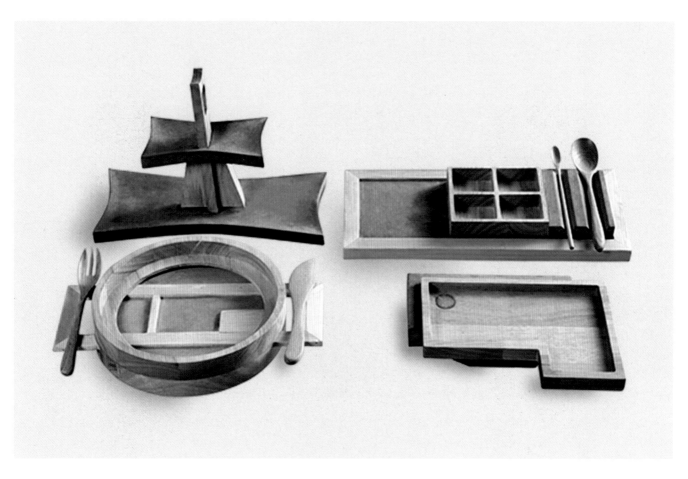

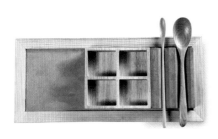

《水泥与木》/ 小食器皿
作者：傅淑萍 颜璐
材质：水泥、木材

Cement and Wood / Small Tableware Series
Designers: Fu Shuping, Yan Lu
Materials: Cement, Wood

　　混凝土是一种坚硬的材料,但是在浇注凝固前是流动和可塑的。该设计选取木与混凝土结合,赋予器物一种几何学秩序,追求材料的本质运用,展现出敦厚、朴质、细腻的材料特质。

Cement is a tough material which is mobile and ductile before casting concretion. In the design, wood and cement are combined to endow a geometric order to ware, the essence of material would be fully applied to reveal the simple, natural, delicate characters of material.

《四季》/ 纸胎漆器
作者：邓莲 王立端
材质：纸、植物漆

Four Seasons / Paper Body Lacquer Ware
Designers: Deng Lian, Wang Liduan
Materials: Paper, Lacquer

传统漆艺极能体现"包容"二字，因材料无以高低贵贱、软硬兼施均可入漆。作者在传统脱胎漆器工艺基础上，利用廉价手工宣纸的绵软特点制作纸胎造型及形态肌理，在漆的固化下形成坚实的器物，并根据其肌理施以漆色来表达自然形态，寓意四季更迭。

The traditional lacquer art embodies the meaning of "inclusion" as all material, regardless of expensive or cheap, soft or hard, could be used in lacquer. Based on the fundamental process of traditional bodiless lacquer, the author utilize the soft character of cheap rice paper to form the outline and appearance texture of paper body, then generate firm ware after solidified by lacquer, then use the lacquer color in the texture to convey the natural features with the implied meaning of season alternation.

《意陶二十四器》/ 气体成型陶瓷
作者：尧波
材质：陶

"24" Pottery / Gas-molding Pottery
Designer: Yao Bo
Material: Pottery

中国饮食器物设计作品集　Design of Chinese Dining Utensils | 33

　　作品《意陶二十四器》是对 2013 年 Jonathan Crary 的精彩著作《24/7》的回应。Crary 从睡眠、工作的角度分析了当代全球资本主义系统的时间感如何加重了当代人的困境。生成不是一种被量化的时间，它与宇宙、自然的节奏息息相关，例如中国农历的 24 节气。作者实验在两种材料（泥浆、气球）之间的运作，利用气体进行器物造型，让火的痕迹在气体升华中留存于器皿之上。这是一种关于偶然性的思考方式和生活方式，它既是制陶者在日常生活中的艺术实践，也是探讨未来的一种方式。

The work "24" Pottery is the response to the splendid work <24/7> by Jonathan Crary in 2013.Crary analyzed how the sense of time in modern global capitalism system aggravate the dilemma of modern people from the aspect of sleep-work. An un-quantized time which is closely bound up with the rhythm of universe-nature is formed, such as 24 solar terms in Chinese lunar calendar. The author perform the experiment with two materials (slurry-balloon), use air to form the outline of vessel, keep the trace of fire in the ware in the course of sublimation. This is a way of thinking regarding the contingency and way of living, it is the way the pottery maker perform artistic practice in daily life and explore the future.

民以食为天 Food is God | 34

《意陶二十四器》/ 气体成型陶瓷
作者：尧波
材质：陶

"24" Pottery / Gas-molding Pottery
Designer: Yao Bo
Material: Pottery

他山之石，可以攻玉。借助现代新型材料作为传统漆艺造型的工具，胎体能够快速成型，且漆易于附着其上、质地轻盈。作品尝试回溯远古时期人类凿石为器的自然造物精神，虽然状似顽石，实则轻若无物，粗糙的表面肌理与内部的金银光泽形成强烈的视觉对比，隐喻拙朴事物的内在精神价值。

"Stones from other hills may serve to polish the jade of this one." This saying means "others' experiences can be lessons." The modern new materials are adopted as tools for the traditional lacquer art. The base can be quickly molded, and the lacquer can be easily attached onto it. The whole texture is light and graceful. The whole work attempts to explore the spirit of humans adopting stones as tools to create items for their use in the ancient times. Though it looks like a hard rock, it is in fact light and without anything within. The coarse texture and the internal gold and silver luster form a strong visual contrast, which implies the internal spiritual value of plain things.

《他山之石》
作者：李瀚然　蒲柯宇
材质：纸黏土、大漆

Stones From Other Hills
Designers: Li Hanran, Pu Keyu
Materials: Paper clay, Lacquer

《相逢》/ 铸铁茶具
作者：周亚蓬　王立端　吴菡晗
材质：铁、铜、木

Meeting / Cast Iron Tea Sets
Designers: Zhou Yapeng, Wang Liduan, Wu Hanhan
Materials: Iron, Copper, Wood

作品从材料、造型、文化三点出发，在材质上让自然细腻的木质与硬朗粗糙的金属"相逢"；造型上以锋利硬朗的线条与动感的曲线结合，形成曲与直形式上的"相逢"；文化理念则是东方气韵精神与西方理性思维的"相逢"。同时，不同材质的结合既使器皿具有更加丰富的视觉效果，同时又解决了传统铁壶金属把手过热的问题，实现了材质美与实用性的和谐统一。

The work design is revealed in the aspects of material, outline and culture, in the aspect of material, the delicate wood meets with tough metal; in the aspect of outline, the sharp line is combined with dynamic curve which enable the curve meet with straight; in the aspect of culture, the oriental conception meet with occidental rational thinking. Meanwhile, the combination of different material not only brings abundant visual effect to the ware, but also solve the problem of overheated metal handle in traditional iron pot, realized the unity of material aesthetic and practical.

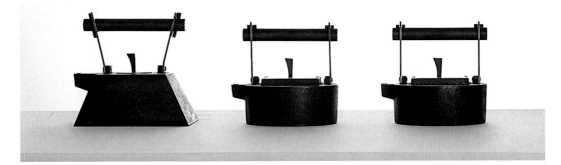

　　漆器是中国最古老的艺术，代表着中国的国粹，属于国家非物质文化遗产，它坚实轻便，耐热耐酸，抗潮抗腐。由于漆器制作费工费时、耗材耗料，因此它仅服务于贵族、文人和宗教，很难在寻常百姓人家看到漆器的踪影。

　　金缕漆器的设计灵感来源于汉代时期的金缕玉衣，因为制作程序复杂，每个玉片都是精雕细作，所以它仅作为皇帝和高级贵族死后的殓服。此系列作品用传统的脱胎漆器制成底胎，用高温让器皿软化后进行切割，再用金线串联缝合而成。金缕漆器作品在历经造型、装饰、完成、破坏、再造的制作过程后，呈现出一种全新的独特效果。

Lacquer is an ancient Chinese art which is the quintessence of Chinese culture, the art is national intangible culture with the features of firm and light, heat resisting and acid resisting, anti-moisture and anti-corrosion. The manufacture of lacquer costs both labor and time, as well as material, therefore the lacquer is only available to nobility, scholar and religion, it is rare to find lacquer ware in the house of common people.

The design inspiration of gold-threaded lacquer ware is originated from the jade suit sewn with gold thread from Han dynasty, the process of the jade suit sewn with gold thread is complicated, with every jade plate curved and threaded by gold, the suit is used for the died emperor and senior nobility. This series of work used bodiless lacquer as bottom body, then use high temperature to soften the vessel for cutting, then sew the pieces with golden thread. The gold-threaded lacquer ware presents a unique expression after the procedure of modeling, decoration, complete, destroy and regeneration.

《再造·重生》/ 金缕漆器系列
作者：何艳 陈静
材质：夏布、铜丝、大漆

Regeneration / Gold-threaded Lacquer Ware Series
Designers: He Yan, Chen Jing
Materials: Grass linen, Copper wire, Lacquer

中国饮食器物设计作品集　Design of Chinese Dining Utensils | 39

作品以蔬果为灵感，运用拟人的手法讲述蔬果与器物的关系，蔬果生长为器物，以多种顽皮的姿态连体组合为花器之体，瓶口则为色彩斑斓的蔬果剖面，构建了一个好玩的铜质世界，传递"食与器"的关系思考。在工艺上运用失蜡浇铸与独创的"引火入水"金属秘色技法，通过对器物形态与材质的多重表达从而呈现有趣的食与器——疯狂蔬果。

Vegetables and fruits are the inspiration of this work. The personification method is adopted to narrate the relationship between vegetables and fruits, and utensils. The vegetables and fruits are growing into utensils. Connected combinations featuring multiple playful postures form the body of vases. The opening of vases features the colorful profile of vegetables and fruits. An interesting copper world is built and the relationship between "food" and "utensil" is reflected. In terms of the technique, the paraffin casting-forcing and the unique "fire to water" metal olive-green technique are adopted. Based on multiple representations of the utensil shapes and the materials, an interesting combination of crazy creatures and vases is finally achieved.

《疯狂蔬果》/ 花器组合
作者：罗黛诗
材质：铜

Crazy Creatures / Vases
Designer: Lou Daishi
Material: Copper

中国传统青瓷之碎片与透明玻璃结合，采用修复、嫁接、延伸、重构等手法，突显两种材料的文化性、艺术性。玻璃似有似无之感衬托了青瓷的存在感和审美性。其造型体现了中国传统饮食文化喜庆的风格，是当代艺术语境下对于传统文化新的诠释。

The design combines the fragments of traditional Chinese celadon and transparent glasses, and adopts methods of repair, grafting, extension and reconfiguration to highlight the cultural and artistic quality of two kinds of materials. The indistinct feeling of glass sets off the existence and aesthetic nature of celadon. The model shows the festive style of traditional Chinese food industry, and is the new interpretation of traditional culture under modern art environment.

《青渺》/ 玻陶器皿系列
作者：范易 贾倩 郭桂林 李尧瑶 袁雁辉
材质：玻璃、陶瓷、漆、铜钉

"Celadon Charm" / Glass & Pottery Series
Designers: Fan Yi, Jia Qian, Guo Guilin, Li Yaoyao, Yuan Yanhui
Materials: Glass, Ceramics, Lacquer, Copper nail

在石材商品生产的过程中会产生许多石块废料。废弃的原石具有凹凸不平的表面，利用废弃石块的不同特点，直接在原石上制作出器皿。原石凹凸的表面与经过打磨的器皿空间形成对比，仿佛器生于石，而石融为器。

Numerous waste stones would be left in the production procedure of stone products. The wasted rough stone have ragged surfaces; the ware is produced with the rough stone, with fully utilization of the characters of the stone. The contrast between the ragged surface of stone and the polished ware would be explicit, as if the ware is generated from the stone while the stone is integrated with the ware.

《石器》/ 青白玉器皿
作者：罗黛诗 雷霞
材质：大理石

Creative Stone Ware / Light Greenish White Jadeware
Designers: Luo Daishi, Lei Xia
Material: Marble

《新概念泡菜坛》
作者：杨曼羚　邹红嫒　周丽雯
材质：陶瓷、木

Creative Pickle Jar
Designers: Yang Manling, Zou Hongyuan, Zhou Liwen
Materials: Ceramics, Wood

泡菜本为冬日延长蔬菜食用期限的腌制方法,在中国人的生活和饮食文化中有着极其重要的地位,几乎家家做、人人吃。即使在当代,腌制泡菜也是中国家庭不可或缺的一种生活方式。

本设计提取中国传统泡菜腌制的古老智慧,将古法泡菜的腌制器具进行改良,保留传统的密封方式,对泡菜坛体积大、占用空间、不易取菜、不易搬运的问题进行再设计,使之适应当下家庭人员少、节奏快、住房紧凑的居家环境。

Pickle is a method to prolong the preservation period of vegetables in Winter, pickle is of great significance in Chinese living and food culture, almost every family would make pickles and everyone would eat pickles. Pickle is still an indispensable life style in Chinese family even now.

The design obtains the antique philosophy of Chinese traditional pickle and improves the pickle ware for traditional pickle, reserves the traditional sealing method and designed regarding the problems result from large size of pickle jar, occupy space, inconvenient to take pickle out and handle the jar. The design enables the jar suitable for the present living conditions of less family members, faster living tempo, and more compact living space.

《新概念泡菜坛》
作者：杨曼羚　邹红媛　周丽雯
材质：陶瓷、木

Creative Pickle Jar
Designers: Yang Manling, Zou Hongyuan, Zhou Liwen
Materials: Ceramics, Wood

该设计打破传统陶瓷器型对称、造型工整、千器一面的常态，使用软质纤维材料塑型，经喷浆制作出陶胎，经素烧后上釉，二次复烧制成。器物具有壳薄生动，各不雷同，件件均为单品的特点。

The design of thin pottery breaks the norm of traditional pottery featured by symmetrical shape, monotonous appearance and modeling neat, it is produced by using soft plastic fiber material, producing pottery tires through spraying and then putting unglazed pottery into the secondary complex. This objective is characterized by its vivid thin shell and unique face.

《馄饨皮》/ 薄胎陶瓷器皿
作者：夏扬扬
材质：陶瓷、漆

Wonton skin / Thin pottery
Designer: Xia Yangyang
Materials: Ceramic, Lacquer

《山海杯》
作者：王立端 白晓宇 范易 何源源 张瀚文 陈梦秋 廖桦 任宇 李瀚然
材质：玻璃、陶、木、大漆、瓷、纸、金、银、铜、竹、砂

Mountain and Sea Cup
Designers: Wang Liduan, Bai Xiaoyu, Fan Yi, He Yuanyuan, Zhang Hanwen,
　　　　　　Chen Mengqiu, Liao Hua, Ren Yu, Li Hanran
Materials: Glass, Pottery, Wood, Chinese lacquer, Porcelain,
　　　　　　Paper, Gold, Silver, Copper, Bamboo, Sand

"杯不在大,致力聚积山川江河湖海之灵气;形不在繁,专修精炼金木水火土之精华。"一个设计团体的集体参与,各式各样的材料构成,丰富精细的工艺,塑造出18个造型相似而简洁的"山海杯"。它们或朴实,或清透,或炫目的视觉效果仿佛一个五光十色的山海世界。同时,如此这般的集体合作凝聚了团队的精神力量,活跃了团队的创作氛围。

"The cup is not cup, but it gathers the anima of mountains, rivers, lakes and oceans. The cup does not has a complex shape, but it refines the essence of fire, water, wood, gold and soil." This work is an outcome of collective efforts. Various materials and refined process help shape 18 "Mountain and Sea Cups" with the similar and concise shape. They are either plain or transparent or dazzling. A colorful world of mountains and seas is thus created. The teamwork enlivens the creation atmosphere of the team, and inspires the spiritual force.

民以食为天 Food is God | 48

《草草成器》
作者：王立端 叶凌杉 吴菡晗
材质：稻草类植物、糯米、大漆

Folk Wisdom
Designers: Wang Liduan, Ye Linshan, Wu Hanhan
Materials: Straw, Glutinous rice, Lacquer

中国作为历史悠久的农耕文明国家，自古就有利用天然生物材料的传统。该套器物以稻草与糯米粉为材料，通过传统手脱模方式造型，再施以天然生物漆料，经打磨工艺使器物呈现出天然偶成的特殊肌理。如何挖掘传统生活方式中的民间智慧，运用设计和传统文化的力量推动社会可持续发展，这是设计界应积极思考和努力实践的课题。

China, as a long history of agricultural civilization country, has a tradition of using natural biological materials. The set of objects using straw and glutinous rice powder as materials, modeling by way of traditional hand mold releasing, and then subjected to natural biological paints by grinding process, present a special texture that is naturally accidental into. How to tap the folk wisdom from traditional way of life and use the power of the design and the traditional culture to promote sustainable development of society is the subject of design community should strive to think positively and practice hardly.

民以食为天 Food is God | 50

器节通"气节",器,指器皿,节,指竹节。"君当如竹,坚韧不拔显气节。"竹在中国的传统文化中被誉为君子气节的象征。设计试图通过工艺与材料传达作品所承载的精神内涵:千锤百炼的"锤起铜器"手工艺,代表竹的坚韧精神与正直气节;木材质谦和的形象表达竹的虚心品质。同时,在造型方面,设计融合了竹与竹节的形态与上下组合功能,实现了造型与功能、材质与文化的融合统一。

"Qijie" is a pun, which also means "integrity" (also pronounced as "Qijie" in Chinese). "Qi" means "ware", while "Jie" means "bamboo joints". It is said "a gentleman should be like a bamboo, which is tough and upright." In traditional Chinese culture, bamboo is a symbol of the gentleman's integrity. This design attempts to convey the spiritual connotation using the traditional process and materials. The refined "hammered copper ware" stands for the tenaciousness and uprightness of bamboos. The modest image of wood shows the modesty of the bamboo. In terms of shaping, the design integrates the shape of bamboo and bamboo joint, and their functions, thus realizing unity of shape and functions, materials and culture.

《器节》/ 锤起铜器
作者:何源源
材质:铜、木

Qijie / Hammered Copper Ware
Designer: He Yuanyuan
Materials: Copper, Wood

"皴"是中国绘画的一种传统技法，常用于表现自然物象的肌理形态，如中国山水画中的"山石、峰峦、树皮"常采用这种技法。作品借用中国画皴法重塑铜器表面的肌理，探索与挖掘铜材质的艺术表现形式，表现中国绘画的艺术与审美价值。作品结合传统漆器工艺，通过对比与反衬表达出两种材质的视觉丰富性和艺术魅力。

Cun is a traditional method for Chinese painting. It shows the shades and texture of rocks and mountains by light ink strokes in traditional Chinese landscape painting. The method is usually used to paint mountains, peaks and barbs in Chinese landscape painting. This work adopts Cun to reshape the surface texture of the copper ware, explore the artistic expression mode of copper materials, and express the art and aesthetic value of Chinese paintings. The traditional lacquerware process is also combined. The contrast shows the diversified visual effect and the unique artistic charm of the two different materials.

《皴》/ 铜胎漆器
作者：何源源 王立端 黄欣
材质：铜、纸、植物漆

Cun / A Copper Base Lacquer Ware
Designers: He Yuanyuan, Wang Liduan, Huang Xin
Materials: Copper, Paper, Plant paint

民以食为天 Food is God | 52

《晴雪、家园》/ 漆艺食器系列
作者：张国栋　张津亚
材质：夏布、大漆

"Sunny & Snow, Homeland" / Lacquer Tableware Series
Designers: Zhang Guodong, Zhang Jinya
Materials: Grass linen, Lacquer

利用传统脱胎工艺技法，在造型上展现脱胎漆器深浅不一的沟壑肌理，通过漆髹饰渲染光的效果，体现出漆艺作品的魅力。红蓝两组髹饰颜色系列，既如火焰的翩翩叠影，又似星空的奇妙玄幻。

The design adopts traditional bodiless process technologies to show the irregular gully texture of bodiless lacquer wares in the aspect of modeling, and show the glamour of lacquer works is shown by light rendering by paint. The red and blue painting ornaments like dancing flames or wonderful starry night.

《砂器》/ 荥经砂器器皿
作者：廖桦
材质：砂、泥、木材

Grit Utensils / Yingjing Stoneware
Designer: Liao Hua
Materials: Sand, Mud, Wood

中国四川的荥经砂器具有耐高温、透气等特点，一直是中国民间的理想餐饮具。本系列产品根据砂器自身特点，通过设计将传统砂器与原木材质相结合，既保留了砂器的传统功能、古朴造型，同时融合当代审美理念，从而让传统手作具备适合现代家居环境使用的气质和功能。

The Yingjing stoneware in Sichuan Province, China has characteristics of high temperature resistance and ventilation, and has always been the ideal folk tableware in China. The series of products, according to characteristics of stoneware, combines traditional stoneware and log materials, which not only reserves the traditional function and simple model of stoneware, but also integrates modern aesthetic ideas so as to make traditional work with quality and function suitable for modern home environment.

《无题》/ 纸胎漆器
作者：邓莲
材质：纸、植物漆

Untitled / Paper Body Lacquer Ware
Designer: Deng Lian
Materials: Paper, Lacquer

传统与现代不只意味着打破与重塑，同时也可以兼容并蓄、相互依存。作者在制作此系列纸胎漆器的过程中，试图用传统漆器中的红、黑、银表现漆器的固有风格，也尝试用多样的色彩和不平的表面来探索纸胎漆器的更多可能性。

The integration of tradition and modern does not only imply break through and remodeling, but also could be inclusive and mutually interdepended. In the production process of this series of paper body lacquer design, the author intended to use red, black and silver in traditional lacquer design to express the inherent style of lacquer and used various color and uneven surface to explore the possibilities in paper body lacquer design.

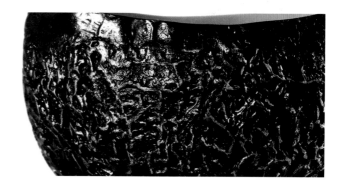

纤细的竹丝，经纬交织，紧扣瓷胎，依胎成形，和瓷胎浑然一体。该套作品灵感来自"君子竹"，寓意君子高风亮节的品德与气节，亦有节节高升的美好祝愿。以洁白如玉的景德镇瓷器为胎，竹编质地坚韧，劈成薄如绸绢，细如纱线的篾丝。白瓷青竹，泥竹一世界。

The slender bamboo filaments cross with each other and cling tightly to the porcelain base. The bamboo filaments and the porcelain base are integrated with each other. The work gets its inspiration from "Junzizhu" (literary meaning "gentleman bamboo"). It implies the exemplary conduct and nobility of character of a gentleman, and also wishes for continuous improvement. The white Jingdezhen porcelain is adopted as the base, the bamboo-woven texture is firm and is split into silk-like thickness. The white porcelain and the green bamboo filaments create a world of their own.

《竹节·丝语》/ 竹丝瓷胎器皿
作者：任宇 谢睿
材质：陶瓷、竹

Bamboo Joints and Filaments / Bamboo filament porcelain base utensils
Designers: Ren Yu, Xie Rui
Materials: Pottery, Bamboos

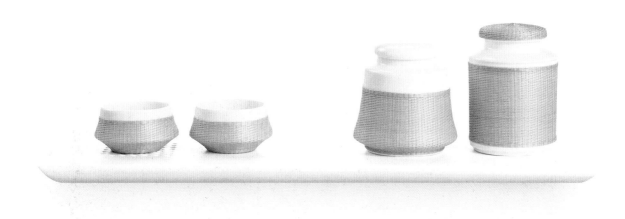

竹丝瓷胎是巴渝地区独有的汉族传统手工艺,白瓷器皿为胎,以纤细如发的竹丝、柔软如绸的竹篾依胎编织而成。竹丝游走在温润如玉的白瓷表面,在编织过程中,瓷胎的轮廓变幻,线条的动静呼应,是竹丝编织独特的肌理,也是淡雅清新的东方气质与文化。

The bamboo filament porcelain base is a traditional handicraft of the Han People in Bayu Region. The white porcelain is adopted as the base. The sender bamboo filaments and the soft and silky bamboo splits are used to weave along the porcelain base. The bamboo filaments surround the mild white porcelain. During the weaving process, the profile of the porcelain base changes. The static and dynamic lines reinforce with each other. The unique texture woven by bamboo filaments shows the refreshing Oriental temperament and culture.

《咏竹》/ 竹丝瓷胎器皿
作者:任宇 谢睿
材质:陶瓷、竹

An Ode to Bamboos / Bamboo filament porcelain base utensils
Designers: Ren Yu, Xie Rui
Materials: Pottery, Bamboos

《旋切木艺》
作者：吴时敏
材质：木

Rotary-cutt Wood Art
Designer: Wu Shimin
Material: Wood

按常规旋切制作的器物造型是规整的，这既是其优势又是其局限。如何能突破规整? 这是一种挑战。打破思维定势,改变对木头的夹持方式和切削方式,保留了更多木头原始自然结构和肌理,使作品更富于变化并获得更强烈的原始趣味和人情味,加强了作品的视觉与触觉美感。木头从材到器、从器到道, 完成了从物质性到精神性的升华。

Utensils made according to the conventional cutting technique are well structured. This can be an advantage or a disadvantage. How should these utensils make a breakthrough of their structure? This is a challenge. The renewal of the fixed thinking can change the mode of wood clamping and cutting, and retain the natural structure and texture of the wood. In this way, the work can be full of changes, and achieves more fun and human interest. The visual and touch aesthetics of the work can also be enhanced. From materials to utensils and from utensils to Tao, the wood finishes the sublimation from the material level to the spiritual level.

面点是中国烹饪的主要组成部分，素以制作精良、品类丰富、风味多样著称于世。此作品采用琉璃之艺术手法，通过光影塑造面点精致玲珑的质感，让观者有一种欲食欲看的时空交错感。

As a major component of Chinese cuisine, pastry is known for its well-made techniques, rich categories, and diverse flavor to the world. This work, through adopting artistic glazing method and using light shadow to shape delicate and exquisite pastry, delivers a magic and fantastic sense to viewer and appeal to their appetite.

《玲珑面点》
作者：范易　贾倩　郭桂林　李尧瑶　袁雁辉
材质：玻璃

Delicate Pastry
Designers: Fan Yi, Jia Qian, Guo Guilin, Li Yaoyao, Yuan Yanhui
Material: Glass

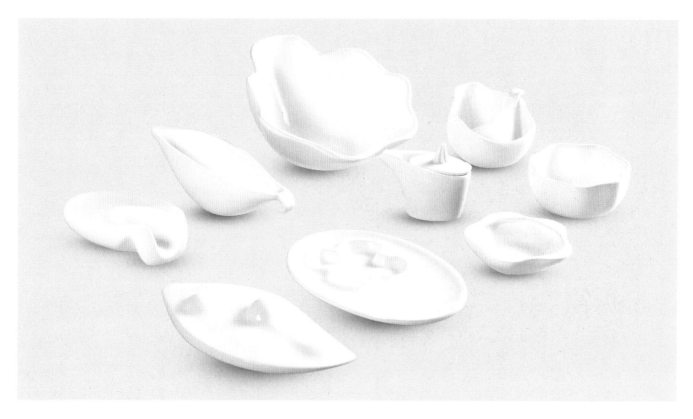

端午节是中国传统节日之一,已被列入世界非物质文化遗产。节日期间,人们皆有吃粽子、喝雄黄酒、食五黄等端午时食的习俗。"时食餐具/端午宴"餐具器皿的设计概念,正是来源于这一传统文化。

作者针对粽子、雄黄酒、咸鸭蛋等"端午时食"盛放及饮食的行为方式,以"花"为碗钵,以"叶"为碟盘,设计"时食餐具/端午宴"器皿,从侧面反映了中国传统节日饮食文化的内涵。

Dragon Boat Festival, which has been included in world intangible cultural heritage, is one of the traditional festivals in China. The custom of Dragon Boat Festival is enjoying seasonal food such as eat rice dumplings, drink realgar wine, eat five kinds of food with yellow color. The design concept of ware in "seasonal food table ware / Dragon Boat Festival Banquet" is originated exactly from this traditional culture.

Considering the place and eating manners of seasonal food for "Dragon Boat Festival", for example rice dumplings, realgar wine, salted duck egg, in the design of ware in "seasonal food table ware / Dragon Boat Festival Banquet", the author designed the bowl with the image of "flower", and dish with "leaf", the design indirectly reflect the connotation of Chinese traditional festival food culture.

《时食餐具》/ 端午宴陶瓷系列
作者:赵卫东 彭科星 刘倩娇 马林明
材质:陶瓷

Seasonal Food Tableware / Dragon Boat Festival Banquet
Designers: Zhao Weidong, Peng Kexing, Liu Jiaoqian, Ma Linming
Material: Ceramics

这组器物主要的材料为夏布,夏布是一种历史悠久的传统手工艺,以苎麻为原料编织而成,有"天然纤维之王""中国草"的美称,一般用于服装和工艺品。以未经精制、质地生硬、颜色微黄的夏布为生布制成的器皿,既有器物所需要的硬度,又有布料所带来的柔软,两者结合在一起,相得益彰。

作品灵感来源于夏末枯萎的荷叶,颜色形态和夏布类似。这组作品包括4个碟子、3个盘子和1个纸巾筒,主要盛放四川传统小吃叶儿粑、蒸饺、小笼包等食品。

The main material for this group of wares is grass linen; as a historic traditional handicraft, grass linen, which is commonly used in fashion and art work design, is weaved by ramie which has the reputations of "king of natural fiber", "China grass". The grass linen without refine is stiff in texture with faint yellow color and called as raw linen, raw linen could bring both the stiffness of ware and the softness of cloth, the two characters are combined and complemented each other in ware production.

The design is originated from lotus leaf weathered in late summer which is similar with grass linen in color and shape. There are 4 dishes, 3 plates and 1 tissue canister in this work, the ware is mainly used to place Sichuan traditional food such as leaf wrapped rice, steamed dumpling, steamed buns.

《夏布食器系列》
作者:白晓宇 陈石
材质:夏布、大漆

Grass Linen Tableware Series
Designers: Bai Xiaoyu, Chen Shi
Materials: Grass linen, Lacquer

这组作品主要是果盘，充分发挥了夏布织物的特点，做成一组类似鲜花怒放的果盘。

The characters of grass linen have been fully utilized in this work, the material is designed as a fruit tray in the shape of blooming flowers.

这组作品主要利用夏布的硬朗形态和折纸的工艺，最后做成储物罐，因为主要是为了放置花生、红枣等食物，所以命名为《秋实》。

The tough form of grass linen and paper folding craft are used in this work, the work is a storage jar the usage of which is to store peanuts, red dates as well as other food, therefore the work is named as "Autumn Fruit".

《竹编焐库》
作者：王晨雨
材质：稻草、竹

Bamboo-woven Insulated Container
Designer: Wang Chenyu
Materials: Straws, Bamboos

焐饭是保存饭菜温度的传统民间技巧,将热饭菜放入稻草编织的容器中封存,以保持2~3小时的温度,待晚归的家人食用。作品在稻草内胆的内外加上竹编,且可以拆卸分层晾晒,既弥补稻草参差不齐的视觉缺陷,又保持器物本身的保温效果,在延续传统民间智慧的基础上加以美化,为家人带来暖胃又暖心的食物。

To put hot rice and dishes in the straw-woven vessel is a traditional technique of keeping food warm for two to three hours. In this way, the home taste can still be kept for those returning home late at night. The work adds bamboo weaving to the inside and outside of the straw liner. The bamboo weaving can be dismounted for airing. This makes up for the visual defects of uneven straws and keeps the insulation effect of the container as well. While inheriting the traditional folk wisdom, the design is also beautified so as to keep warm food for family members coming home late.

民以食为天 Food is God | 68

四川素有"天府之国""蜀中江南"的美誉，拥有独特的饮食文化。四川盛产竹，此设计搭配细腻的陶瓷，将精致与粗犷巧妙结合，对四川小吃的种类进行对应设计，使餐桌上能够别有一番情趣，也让人透过餐具，体味出中国饮食文化的独具匠心。

With unique food culture, Sichuan enjoys the reputation of "Land of Abundance", "Jiangnan area in Shu". Sichuan is abound in bamboo which is matched with delicate ceramics in this design, the delicacy is combined ingeniously with the toughness; the design is correspondence to the variety of Sichuan snacks, which brings a special taste to the dinner table, people can appreciate the originality of Chinese food culture through the ware.

《四川小吃器皿》
作者：曹宇嘉　张婷
材质：陶瓷、麻绳、竹

Sichuan Snacks Ware
Designers: Cao Yujia, Zhang Ting
Materials: Ceramics, Hemp rope, Bamboo

《沙岩》/ 纸胎漆器
作者：王立端
材质：纸、植物漆

Sandstone / Paper Body Lacquer Ware
Designer: Wang Liduan
Materials: Paper, Lacquer

在纸上可以写出华丽的诗篇或画出美丽的图画。其实,我们还可以利用纸来做造型,用漆来进行装饰,以别样的视觉形态表达出不同的肌理质感,抒发情怀。作者希望这组用纸为胎造型的漆艺器皿能够以象征寓意的手法表达出对沙、岩、矿、泥的感悟。

We can write out magnificent poems or draw beautiful painting on papers. Actually, we can use paper for modeling, use lacquer for ornament, and use different visual form to present quality of different textures and express feelings. The designer wishes this group of lacquer ware modeled by paper body can express perception of sand, rock. Ore and mud by symbolic meaning.

《贝·壳》是一次现实与幻想的碰撞,灵感是将自然中具有共同形态的贝与瓜赋予新意结合。优美的弧度,似水的线条,恰好的颜色,使之成为难以挑剔的作品取材。采用夏布作胎,表面进行变涂造漆,最后抛光完成。数层金粉若隐若现,似活在海中安静沉睡的精灵,给人们带来沉稳的力量,激励人们追求梦想,踏实前进。

The work is a collision between reality and fantasy. The inspiration is the combination of shells and melons with the same shape. The elegant rondure, the water-like lines and the precise color match turn them into perfect materials for this work. The grass cloth is adopted as the base, the surface features variable paint, and the polishing is finally adopted to finish the whole work. Numerous layers of golden powder are partly visible and partly invisible. They are like elves speaking in the sea quietly, calming people down and encouraging people to move ahead to pursue their dreams.

《贝·壳》
作者:李肖依 袁星来
材质:夏布、漆

Shells
Designers: Li Xiaoyi, Yuan Xinglai
Materials: Grass cloth, Paint

以在中国有着深厚历史与研究土壤的蚕丝为材料,将柔软的蚕丝塑型成立体的器皿,再通过自身的张力在钢筋上缠绕,一刚一柔的结合,萦绕出一种虚实相生的空间感。通过反思传统文化中物质和精神的馈赠,还给自己一个似曾相识却陌生的身体经验。作品出于传统,归于一种既定的心手相应的秩序。

Natural silk with a profound history and research background in China is adopted as the material. The soft natural silk is shaped into a three-dimensional utensil. Then, it twists on the reinforcing steel bar with its inherent tension. The combination of stiffness and gentleness creates a sense of space combining fantasy and reality. By reflecting on the material and spiritual relics of traditional culture, I create a strange but seemingly familiar physical experience. The work comes from the tradition and returns to a spontaneous order.

《缱》
作者:蒲柯宇
材料:蚕丝、清水丝绵、金属

Tie
Designer: Pu Keyu
Materials: Raw silk, Qingshui silk floss, Metal

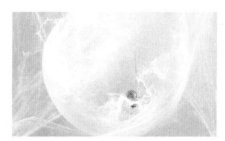

该设计灵感来源于中国传统的包袱,利用花布包袱的特点,将其使用方式及外形运用于食盒的设计当中。设计根据装置物品的实际需要,对食盒每层结构布局进行了针对性分割,并由此让食盒四周各出现了一条斜线。通过有意识的设计打破了传统的食盒内部和外部的固有形式,从而融合传统与现代,创造了具有一定新意的食盒。

The design inspiration is originated from the traditional cloth-wrapper in China; the design utilizes the character of figured cloth, the usage as well as the outline of which have been applied in the design of hamper. In the design, the structural layout of each layer of hamper has been separated subject to the specific requirement of goods to be placed, a slash is generated in each side of the hamper. The design intentionally break through the existing internal and external format of traditional hamper, consequently present an innovative hamper which combines the traditional and modern design.

《食盒(一)》
作者:何欢 张建平 罗显怡
材质:木、漆

Hamper I
Designers: He Huan, Zhang Jianping, Luo Xianyi
Materials: Wood, Lacquer

中国饮食器物设计作品集　Design of Chinese Dining Utensils　| 75

作品以中国传统包袱作为元素，运用大漆工艺，塑造时尚简约的整体造型。结构上，采用嵌入的方式，提手底部与盒身紧紧相扣；装饰上，采用传统工艺贴金箔技法形成极简的几何装饰图案，聚散离合，体现出物与物之间的联系，从而带给观者不一样的认识与视觉体验。

The work adopts the traditional Chinese package as the element. The Chinese lacquer technique is used to mold a fashion and simplified overall shape. Structurally, embedding is employed to closely combine the base of the handle and the box body. In terms of decorations, the traditional gilding technique is used to form extremely simplified geometric decoration graphs. The separation and integration reflects the correlation between objects, and brings different visual and cognition for viewers.

《食盒（二）》
作者：何欢
材质：木、漆、金箔

Hamper Ⅱ
Designer: He Huan
Materials: Wood, Lacquer, Old foil

民以食为天 Food is God | 76

　　作品以中国元素为主题，诠释了中华文化的博大精深，渊远流长。其造型承载了悠久的中国历史文化，也体现了一定的实用性。创作过程中将不同形状的碎木块拼接成型，并采用了变涂、彩绘、戗金、镶嵌等七种装饰技法进行装饰，不仅体现了其工艺难度，更使器皿增添观赏价值。

This work gives full expression to profoundness of the Chinese culture using Chinese elements. Not only has it inherited the profound history and culture of China, but also it is practical. During the creation process, broken blocks of different shapes are pieced together. Seven decoration techniques, including variable painting, colorful painting, gold inlaying and embedding, are adopted. This, on the one hand, increases its technical difficulty. On the other hand, the whole work has more ornamental value.

《漆皿》
作者：陈静　白静　李双利　伍姝梅
材质：漆

Lacquerware
Designers: Chen Jing, Bai Jing, Li Shuangli, Wu Shumei
Material: Paint

该设计颠覆了"漆器必然非常精致"的传统印象，器皿以朴实的原木树枝作为漆器的胎体。因木得形，不加雕琢；材料常见，随处可得。同时，也毫不避讳器物上的自然裂口，使漆器的精致与胎体的拙朴相辅相成。使用者手持器物，一种亲近自然的情趣感觉就会油然而生。

The design overturns the traditional impression that "lacquer ware must be exquisite". The wares use natural rough branches as the body of the lacquer ware. The model is shaped due to the form of the wood shape without sculpturing; the materials are common and can be got everywhere. The ware does not taboo the natural cracks. The delicacy of lacquer ware and the simplicity and nature of body supplement each other so that when the users hold the ware, they feel close to nature spontaneously.

《凿木为器》
作者：王立端　李其龙　糜思尧　贺杰
材质：原木、大漆

Rough Wood Lacquer Ware
Designers: Wang Liduan, Li Qilong, Mei Siyao, He Jie
Materials: Wood, Lacquer

民以食为天 Food is God | 78

  漆与土是最为原始的自然产物，通过古人的智慧得以运用，制成之物更是与人们的生活息息相关。每个器物都有生命，从被制作到被使用，难免磕磕碰碰，就好像一个人的生命历程。漆缮的本意在于，面对不完美的事物，用近乎完美的手段来对待。漆缮代表一种态度，用世上最贵重的物质与精神来面对缺陷，精心修缮，在无常的世界中恪守心中那份对美的向往，化残缺为美，由此获得升华。

Paint and soil are primitive products of the nature. The ancient people skillfully used them to make utensils closely related to daily life. Every utensil has life. From being made to being used, it cannot avoid damage. This is just like how a person grows up. The purpose of paint repair is to achieve perfection of something imperfect. It is a kind of attitude, which uses the most precious thing and spirit to cope with defects. In the ever-changing world, paint repair stays true to its pursuit of beauty. It can turn fragments into beauty and achieve sublimation therefrom.

《惜之，缮之》
作者：张国栋 张津亚
材质：陶瓷、大漆、金

Cherish It, Repair It
Designers: Zhang Guodong, Zhang Jinya
Materials: Pottery, Lacquer, Gold

"谁知盘中餐，粒粒皆辛苦"，唐诗《锄禾》中广为流传的两句，意为餐盘中的粮食皆为辛苦劳动所得，来之不易，用以告诫人们珍惜粮食，切勿浪费的道理。这组器物选用925纯银镀24K黄金，嵌以宝石和珍珠。勺子中盛着的珍珠正是餐盘中"粮食"的象征，表现米之金贵，以金银器皿盛于其中而不为过。作品呼吁人们珍惜一米一粟，传承中华民族勤俭的传统美德。

"Who knows that in your plate; Every grain costs a bead of sweat." These are two famous lines from a poem of the Tang Dynasty, Weeding the young cereal. It teaches people the lesson of valuing every grain and not to waste them. The set of utensils are created with 925 silver-gilt 24K gold and embedded with pearls and gems. The pearls on the spoon are a symbol of "grains" on the dish. It symbolizes preciousness of grains. The set of utensils advocates people to cherish every grain and inherit the traditional national virtue of frugality.

《粒粒皆辛苦》
作者：齐敏达
材质：925纯银镀24K金、珍珠、天然水晶

Every Grain Comes From Hard Working
Designer: Qi Minda
Materials: 925 pure silver-gilt 24K gold, Pearls, Natural crystals

民以食为天 Food is God | 80

　　简化山形与几何方形结合,作品上部为漆器传统工艺技法,下部为棕丝与漆的结合,是豪放与婉约的糅合,两者格格不入而又能有机地结合,碰撞出不一样的味道。一层一层地覆盖,每一层都有颜色和形式的变化,形成镂空层层叠叠的通透感。作品中漆与棕丝融合、重叠、交错,每一层不同的工艺积累,呈现出漆与棕丝的共同语言。

The combinations between the mountain shape and the geometric shapes are simplified. The upper part of the work features the traditional painting skills. The latter part features the combination between the monofilament and the paint, between boldness and conservativeness. The two seemingly contradictory elements are organically combined to achieve a unique effect. The layer upon layer of covering endows every layer with different color and shape changes. Thus, a hollowed and transparent effect is achieved. In the work, the paint and the monofilament are integrated, overlapped and interwoven. Different skills in different layers show that the paint and the monofilament actually share something in common.

《山影》
作者:梁云云
材质:漆、棕丝

Mountain Shadows
Designer: Liang Yunyun
Materials: Paint, Monofilament

中国饮食器物设计作品集 Design of Chinese Dining Utensils | 81

该作品采用中国传统漆器制作工艺,重新诠释乔治·莫兰迪油画中的静物,以一种对话的方式,将过去与现在链接,而此刻的存在又让作品处于永动的进行时,没有终点,没有完结。作品基于实用为原则的髹饰工艺,遗留创作过程纸的拼接痕迹,呈现出对生活的一种认知态度。通过追求一种极致的方式捕捉简单事物的精髓,探寻平凡行为状态中的深层意识。

This work reinterprets still life in paintings of George Morandi using the traditional Chinese painting technique. It uses a dialogue to connect the past with the present, and the present state allows the work to be ongoing all the time without an ending. The work is based on the practical painting technique. The paper collage traces during the creation process are left to show a unique attitude towards life. Elaborately, the work captures the essence of simple things and explores the in-depth consciousness of ordinary behaviors.

《莫兰迪的漆物》
作者: 刘利 张媛媛
材质: 漆、纸

Lacquerware of Morandi
Designers: Liu Li, Zhang Yuanyuan
Materials: Paint, Paper

　　暖暖系列砂器源于童年使用暖水瓶的温暖记忆，由银色的水瓶内胆、瓶盖等延伸而出的黑砂家居用品，组成了一幅温暖的家居画面。此系列作品采用了砂器最具代表性的呛釉工艺：无釉砂器在烧制后，将处于高温状态的砂器放入还原坑中，倒入大量锯木面，锯木面在高温下迅速碳化，附着在砂器表面，形成独特的黑色泛银光的质感。

The warmth series sand-fired utensils can remind one of the warm childhood memories of using thermos bottles. The black sand-fired household items extended from the silver water bottle liner and bottle cap form a warm home furnishing picture. The product series adopts the representative choked glaze process, that is, to put the high-temperature sand-fired utensils to the reduction pit and pour a large amount of sawmilling surfaces. The sawmilling surfaces will be quickly carbonized at a high temperature and attached to the surface of the sand-fired utensils to form a unique black texture with a silver light.

《暖暖》
作者：廖桦
材质：砂、泥、木材

Warmth
Designer: Liao Hua
Materials: Sand, Mud, Wood

重庆荣昌安陶是中国四大名陶之一，其最具有代表性的产品就是泡菜坛子。但随着生活节奏的加快，这种传统生活器物使用得越来越少。所以在设计新的安陶产品时，把经典的泡菜坛子形象与现代生活方式相结合，制作了一套组合餐具。餐具既能在使用时满足人们吃饭的基本需求，又可在叠放时还原为泡菜坛子的形态。

Anfu Pottery is made in Anfu Town, Rongchang County, southwest China's Chongqing. Together with Yixing Pottery in Jiangsu, Qinzhou Pottery in Guangxi, and Jianshui Pottery in Yunnan, they are named as four famous types of potteries in China. The most representative Anfu Pottery product is the pickle jar. However, with quickening pace of life, the traditional life utensil has fallen into obscurity. The typical image of the pickle jar is combined with the modern lifestyle to create a set of furniture combination. The furniture can not only carry food, but also resume to the shape of the pickle jar when being stacked up.

《安陶四合碗》
作者：廖桦
材质：陶

Anfu Pottery Quadrangle Bowls
Designer: Liao Hua
Material: Pottery

以山茶花为创作灵感，配合泥片成型技巧，加以自制植物灰釉后成器。山茶花之美，美在柔中带刚，其花谢时尤为惊人，直接整朵坠落。遂取其果敢坚决之神，以简洁形式表述山花盘。其叶如盔甲，坚韧挺拔，故附以坚锐状表达器皿刚毅之气。叶脉纹设计作为盘内装饰。方形器皿则为表达其生长于大山之间而作。天真之器与天然之食相配尤为适宜。

Camellia is the inspiration of this work. The mud shaping together with the self-made ash glaze forms this work. Camellia is beautiful for alternating gentleness with stiffness. When its blossoms fall, they fall as a whole. Thus, this work extracts the resolute spirit from camellia and represents the camellia disk in a simplified form. The leaves are as firm as the helmets. Thus, the leaves are shaped to be firm enough to show the manhood of the utensil. The branching fibers are designed to be the decorations within the disk. The square utensil is made to show its growing environment, which is surrounded by mountains. The natural utensils and the natural foods are a good match.

《灰釉器皿》
作者：张茜
材质：陶瓷

Ash Glaze Utensils
Designer: Zhang Qian
Material: Pottery

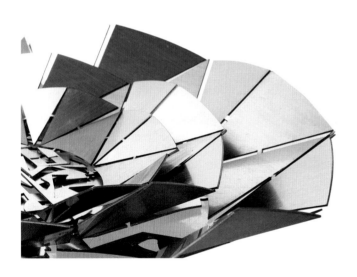

汉字是中国古代饮食文化的折射、反映与记录。在一笔一划中一窥中国饮食文化之奥妙，在一碟一碗中感受中国饮食文化的传承。作品运用现代材料和工艺，将记录饮食文化的文字作为构成器皿的元素，辅以简单的几何线条分割，通过传统与现代的结合与包容，讲述一个个关乎饮食、关乎生活的故事，赋予器物更深层的文化内涵。

Chinese characters are a reflection and record of the traditional Chinese food culture. Every stroke shows the secret of the Chinese food culture. Every dish and bowl is an inheritance of the Chinese food culture. This work uses modern techniques and materials and turn characters recording the food culture into elements of the utensil. Several simple geometric lines are adopted for separation. The combination of traditional and modern elements tells one after another story about food and life, and endows the utensil with deeper cultural connotation.

《食语》
作者：彭威　柳芳
材质：不锈钢

Voice of Food
Designer: Peng Wei, Liu Fang
Material: Stainless steel

The lacquer art is one of traditional Chinese cultural heritages, but it is falling into obscurity. The beauty and value of the lacquer art needs to be carried forward and developed relying on the external attention and protection. By combining the lacquerware and the porcelain, the work expresses the abstract concept of "Space" via the free flow and the expansion of the lacquer tray. The good structure and the stiffness of the porcelain utensils convey a detached and unselfish state. To create a refreshing tea set using the traditional process and culture is not the only purpose of the creator. More importantly, the creator hopes to create a realm of "being detached and unselfish in the space".

漆艺作为中国非物质文化遗产之一，却面临着日益衰微的困境，漆艺之美和价值需藉由获得外界的关注和保护以更好地传承与发展。作品通过漆器与陶瓷的结合，以漆盘自由流动、舒展的造型表达抽象的概念——虚极，用陶瓷茶具规整、坚硬、严谨的特质传达静笃的状态。作者不光是想利用传统工艺文化创作一套新颖的茶具，更想为使用者营造"致虚极，守静笃"的心境。

作品由筷子、筷托、实盘三大主体构成。作品名"箸"：以活字印刷为切入点，融入中国百家姓文化，木、金属两种不同的材质结合设计制作出一款筷子，其顶部为金属反刻百家姓繁体汉字，倒过来即可印出文字，充满了趣味性；"弧"：以金属和木材结合，设计制作出一款筷托，主要用于用餐时临时放置筷子，筷架呈弧形，在施加外力时会左右摇摆，正式中透露着一点俏皮；"纹"：该设计主要采用传统图案为轮廓设计的一款实盘，其边缘过渡平缓圆润，角上小细节为整个作品的亮点。

The work consists of three major parts, namely chopsticks, chopstick tray and food tray. Name "Chopsticks" Chopsticks: This design proceeds from the movable-type printing. The hundred family names in China are integrated. The two different materials, wood and metal, are combined to design the chopsticks. The top part features the metal carving of hundred family names in traditional Chinese. Put it upside down, and the letters can be printed. The whole process is full of fun. "Arc": The metal and the wood are combined to design a chopstick tray. It is used to put chopsticks while having meals. The chopsticks rack features an arc shape. When the external force is imposed onto it, it will vacillate to the left and the right. Being a formal design, it looks smart as well. "Lines": This design adopts the traditional patterns as the profile to design a food tray. The edge is excessively round and smooth. The fine details on the angle add highlights to the whole product.

《虚极与静笃》（上） Detached and Unselfish in the Space
作者：李颖 Designer: Li Yin
材质：亚克力、漆、陶 Materials: Acrylic, Lacquer, Pottery

《方》（下） Square
作者：李其龙 Designer: Li Qilong
材质：木、金属 Materials: Wood, Metal

本设计灵感来自于中国传统建筑构件：梁柱和柱础。特地选用重庆本地陶泥以达到木的质感，对柱础的结构比例进行夸张、提取；同时选用白色开片釉料，使器皿在使用过程中不断变化，历久弥新。

The design inspiration originates from the traditional Chinese building components, including beam columns and column bases. The pottery clay from Chongqing is selected to achieve the wood-like texture. The structural proportion of column bases is exaggerated and extracted. Meanwhile, the white crack glaze is chosen to keep vessels changes and newer with time during the use process.

作品灵感来源于节日茶桌上的瓜果。对于果实的形态进行再创造，并利用不同果实本身的结构分离出底和盖子的形状。用夸张和简单化的方法重组瓜果表面的肌理，总体制造出一种自然朴实的风格。材质：青木关粗陶；制作方式：手捏成型，施无光白釉；烧成方式：1250℃氧化烧成。

The work seeks its inspiration from melons and fruits put on the tea table on festivals. The shape of fruits is recreated. The original structure of different fruits is adopted to separate out the shape of the base and the cover. Exaggerated and simplified methods are used to recombine the texture of fruit and melon surfaces. A natural and plain style is created. Qingmuguang Coarse Pottery is adopted as the material. The manual molding is combined with dark white glaze. Then, it is oxidized and fired at a temperature of 1250°C.

《柱》（上）　　　　　Columns
作者：于瑛豪　　　　　Designer: Yu Yinghao
材质：陶　　　　　　　Material: Pottery

《溯果》（下）　　　　Origin of Fruits
作者：白玥 毛安澜　　　Designer: Bai Yue, Mao Anlan
材质：陶　　　　　　　Material: Pottery

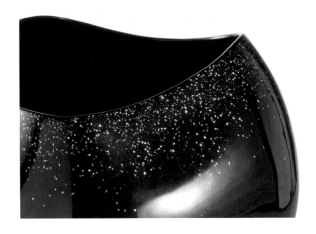

本次的创作定位是日常器物。设计了各不相同、具有石头意趣的造型，作品从每一面看都不一样。器物的瓶口较大，可放置面积大的东西。外部装饰则运用了贴和撒螺钿的髹饰技法，表现出漆器的特有深邃感，通过打磨和抛光展现漆器本身蕴含的温润光泽。

This creation is positioned at daily utensils. Interesting stone shapes of different sizes are created. The work looks different from different perspectives. The bottleneck of the utensil is large enough to hold some large objects. The external decorations employ the painting decoration techniques of pasting and mother-of-pearl inlaying. The lacquerware thus created looks profound. The polishing gives full expression to the luster of the lacquerware itself.

《星石》
作者：温美婷
材质：天然漆、抹布、瓦灰、螺钿

Asteriated Stones
Designer: Wen Meiting
Materials: Natural lacquer, Cleaning cloth, Tile paint, Mother-of-pearl inlay

作品灵感来源于石板岩，作品底胎为纸胎，层层叠加。在创作过程中，融入了新的想法：红色漆器上饰有篆书文字，体现对传统文化的传承；黑色漆器上的装饰是黑红相融，突显光滑与粗糙的对比。同时，制作的似"血"一样从漆器本身流出来的液体质感，有如与漆器一起经历时间雕琢后留下的痕迹。

The inspiration comes from the slabstone rocks. The bottom base of the work features the paper base, which is added layer by layer. During the creation process, new ideas are integrated. The red lacquerware bears the seal characters, which is an attempt to inherit the traditional culture. The decoration on the black lacquerware features the mix of black and red, and highlights the contrast between smoothness and coarseness. Meanwhile, the blood-like liquid flowing from the lacquerware symbolizes the time traces of lacquerware.

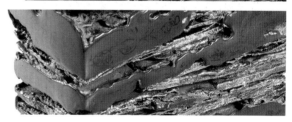

《垣》
作者：冯宇森
材质：漆、瓦灰、纸

Walls
Designer: Feng Yusen
Materials: Paint, Tile paint, Paper

有感于简单朴实的生活沉淀,体味寻常事物带来的感触,作品以普通的抹布为胎,髹以大漆,过程中保留织物的特有形态,或卷曲,或舒展,或堆叠迂回,在寻常无奇的事物中窥探生活中的细微感动。"抹布"在台前幕后的功能角色转变,使日常的一餐、日常的器皿,通过最平凡的抹布得以承载。

A simple life also has touching moments. This work adopts the ordinary cleaning cloth as the base, and is painted by lacquer. During the process, the unique shape of textiles is maintained, being either curly, or spread, or piled up. From ordinary things, the touching moments can be captured. The change of the functional role of "cleaning cloth" helps inherit its most ordinary role—being the carrier of daily meals.

《抹布》
作者:蒲柯宇 李瀚然
材质:布、大漆

Cloth
Designers: Pu Keyu, Li Hanran
Materials: Cloth, Lacquer

　　紫铜器具拥有一种质朴的亲和力,没有金银的豪奢,不似青铜的历史沧桑,以迷人的金属光芒,精致灵动的气质,服务于人们的日常生活。此套紫铜器具一共三件,分别为铜壶、铜杯、铜罐,满足现代人对于茶道的热衷。形态简洁,造型圆润,着色靓丽含蓄,颠覆对传统紫铜器具的印象,服务现代人的审美及使用需求。

The red copper utensils look plain and amiable. They are not so sumptuous like silver and gold, and they don't bear the historical vicissitudes of the bronze. They have a charming metal radiance and an ethereal temperament, and serve people's daily life. The set of red copper utensils consists of three pieces, namely a copper pot, a copper cup and a copper jar. They can meet the passion of modern people for the tea ceremony. Being simple in shape and brilliant in color, the work transforms people's impression of the traditional red copper utensils and caters to modern aesthetic taste and use demands.

《紫金茶具》
作者:陈学渊
材质:紫铜

Red Golden Tea Set
Designer: Chen Xueyuan
Material: Red copper

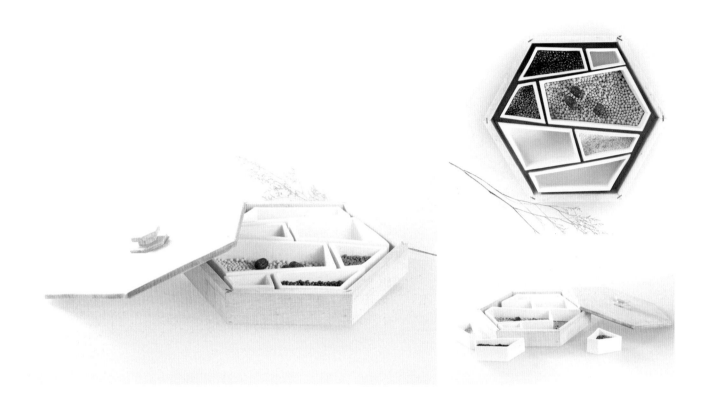

合食两字拆开可理解为"一人一口食"。"合"也有合家、合和、合并之意。作品以传统食盒为设计载体,结合传统苏州园林窗花的设计元素,取合食分食之概念,以重构分割的手法设计而成,旨在传达当下信息化社会与传统饮食方式结合的设计理念。

The two Chinese characters, "合食" (meaning "shared meal" literary), can be divided into "一人一口食" (meaning "one bite one person" literally). "合" also has the meaning of family reunion and harmony. Therefore, this work is created by using the reconstitution and separation design method, adopting the traditional hamper as the design carrier, integrating the design element of the traditional Suzhou garden paper-cuttings for window decoration and referring to the concept of "shared meal". It aims to spread the design concept of combining the current informationized society and the traditional dietary mode.

《合食》
作者:包钰 李洲洲 刘檬 张振环
材质:瓷、大叶楠

Shared Meal
Designers: Bao Yu, Li Zhouzhou, Liu Meng, Zhang Zhenhuan
Materials: Pottery, Machilus kusanoi

泥浆气球成型使大地的肌肤——泥土自行向极限延伸，不断地生成各种形态。这种方式伴随着时间和气候的变化留下不同的渐变肌理，时而悠闲，时而紧迫。这种自行生成的形态在柴窑中经历火焰的洗礼重生成器，转换为日常生活中的平常之物。

The balloon-like mud allows the skin of the land, soil, to extend freely to the limit and generate different forms. The texture changes gradually, being sometimes loose and sometimes compact, along with changes of time and climate. The self-formed shape undergoes the baptism of flames in the wood-fired kiln to form daily utensils.

《肌肤下的火焰》
作者：尧波
材质：陶

Flame under the Skin
Designer: Yao Bo
Material: Pottery

作品以漆器中极其复杂的脱胎手法为基础,第一个系列中选择筷子和竹勺为元素,它们来自中国乡村的传统厨房,以脱胎方式将其显示,第二个系列的肌理来自于厨房的泥土地面,是柴火做饭的传统厨房形式,凹凸不平的地面拓印出特殊而别致的纹理。中国、美食、筷子、饭勺、厨房地面、大漆作为灵感来源,于作者这样在中国的法国人,好像没有联系,却充满故事。

The work is based on the complex technique of making bodiless lacquerware. In the first series, chopsticks and bamboo spoons are adopted as elements. They both come from the traditional kitchen in Chinese villages. The technique of making bodiless lacquerware is adopted to represent the whole work. The texture of the second series originates from the mud land surface of the kitchen. It is a traditional kitchen for cooking with firewood. The rugged ground prints the special and exquisite texture. The second series gets its inspiration from "China, delicacies, chopsticks, spoons, kitchen ground and Chinese lacquer". This seems to have nothing to do with designer, a French living in China, but in fact many of my life stories are related to these elements.

《胎》
作者:文森(法国)
材质:漆

Base
Designer: Vincent Cazeneuve
Material: Lacquer

《清朴·提盒》
作者：张瀚文
材质：黑胡桃木、紫铜

Simplicity Carrier with a Handle
Designer: Zhang Hanwen
Materials: Black walnut, Red copper

精选黑胡桃和紫铜为基础材料。黑胡桃木件通过榫卯工艺结为盒身，盒身之间藉由齿状形面紧致卡合，藏巧于拙，方为大巧。提盒用料厚实，结构牢固，上下双层盒身通过铜件旋钮可自由取合，盒身内部空间充裕，可作为食盒、茶盒等诸多之用。提盒整体器型端庄别致，厚朴苍劲，试以云淡风清之豁达品世间千情百味。

The black walnut and the red copper are adopted as basic materials. The black walnut is used to make the box body via the mortise and tenon joint technique. The box body features the dentated surface and compact clamp. The ingenuity is covered by clumsiness. The whole design looks elegant. The round tiered carrier with a handle is made up of firm materials, and has a firm structure. The up and down double-layered box can be easily closed through the copperware buttons. The box is large and can carry food and tea. The whole carrier looks elegant and forceful. Users can experience all tastes of the world with a generous mind.

民以食为天 Food is God | 98

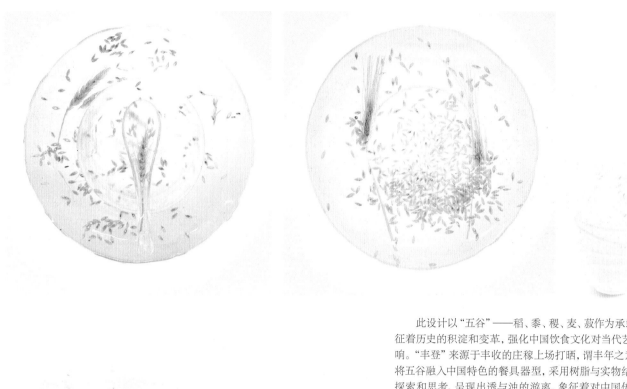

　　此设计以"五谷"——稻、黍、稷、麦、菽作为承载物,象征着历史的积淀和变革,强化中国饮食文化对当代艺术的影响。"丰登"来源于丰收的庄稼上场打晒,谓丰年之意。作品将五谷融入中国特色的餐具器型,采用树脂与实物结合进行探索和思考,呈现出透与浊的游离,象征着对中国传统五谷文化的变迁思考。

The design adopts five cereals—rice, millets, glutinous millets, wheat and beans—as the carrier to symbolize the deposit and reform of history. It also emphasizes the influence of Chinese diet culture on the contemporary culture. A bumper harvest means the busy scenes on the farm. The work integrates the five cereals into the tableware with Chinese characteristics. The resin is combined with the practical items for in-depth exploration and thinking, and to show the separation between transparency and turbidity. It also resembles the reflection on transformation of the traditional Chinese "five cereals" culture.

《五谷丰登》
作者:唐愚程
材质:树脂、稻、黍、稷、麦、菽

A Bumper Harvest
Designer: Tang Yucheng
Materials: Resin, Rice, Millets, Glutinous millets, Wheat, Beans

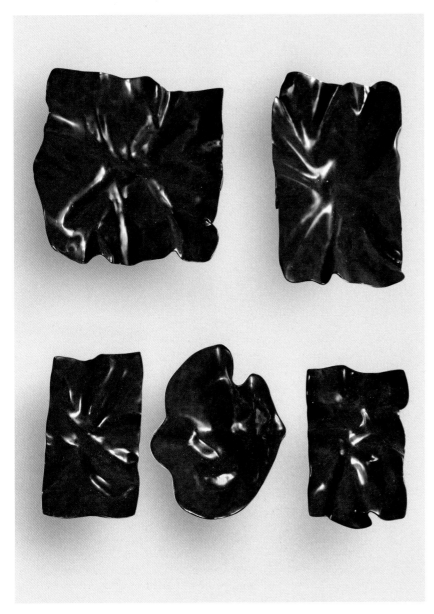

本漆艺作品是用脱胎工艺制作完成。在脱胎过程中完成类似布面皱褶的外形。漆面表现主要以红色调为主，细节上有细微变化。底漆完成后在表面多次戗金，最后达到在柔软的皱褶布面上呈现若隐若现的细纹效果，并让作品能够功能性和美观性兼具。

The lacquer ware is manufactured by bodiless process. The cloth wrinkle like appearance is formed in the course of bodiless process. The lacquer appearance is mainly red, with delicate changes in detail. The surface is inlayed with gold after the bottom lacquer is completed, at last the soft cloth surface would achieve faintly discernible wrinkle which enable the work with the function and the beauty.

《褶皱盘》/ 漆器系列
作者：王远秋
材质：夏布、大漆

Wrinkle Plate / Lacquer Ware Series
Designer: Wang Yuanqiu
Materials: Grass linen, Lacquer

民以食为天 Food is God | 100

作品以四扇家具屏风作为载体，通过中国传统刺绣的方式表达传统节日中具有代表性的美食，主题分别为：元宵节、端午节、中秋节、冬至。每一个节日都有吃相应美食的习俗：元宵节吃汤圆，端午节吃粽子，中秋节吃月饼，冬至吃饺子。作品寄托着中国传统文化中对阖家团圆、新旧交替、吉祥幸福的向往和憧憬。

The work adopts four screens as the carrier, and shows representative delicacies on traditional Chinese festivals using the traditional Chinese embroideries. It has four themes, namely the Festival of Lanterns, the Dragon Boat Festival, the Mid-Autumn Festival and the Winter Solstice. On every festival, Chinese take different delicacies. On the Festival of Lanterns, Chinese have glutinous rice balls; on the Dragon Boat Festival, zongzi (traditional Chinese rice-puddings); on the Mid-Autumn Festival, dumplings. These foods imply yearning of traditional Chinese culture for family reunion, transition from the old year to the New Year, auspiciousness and happiness.

《阖家—锦绣食》
作者：张一璠 李玥 牟彦宣 廖雯霏
材质：绢、蚕丝、实木

The Whole Family – Numerous Delicacies
Designers: Zhang Yifan, Li Yue, Mou Yanxuan, Liao Wenfei
Materials: Thin, Tough silk, Natural silk, Solid wood

中国饮食器物设计作品集 Design of Chinese Dining Utensils | 101

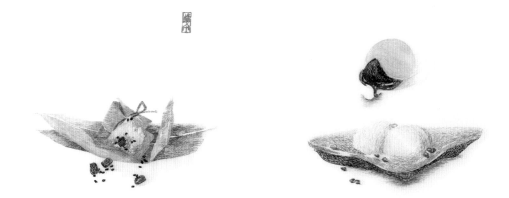

《净桌》/ 木八仙桌
作者：吴菡晗
材质：黑胡桃木

Clean Table / Wood Square Table
Designer: Wu Hanhan
Material: Wood

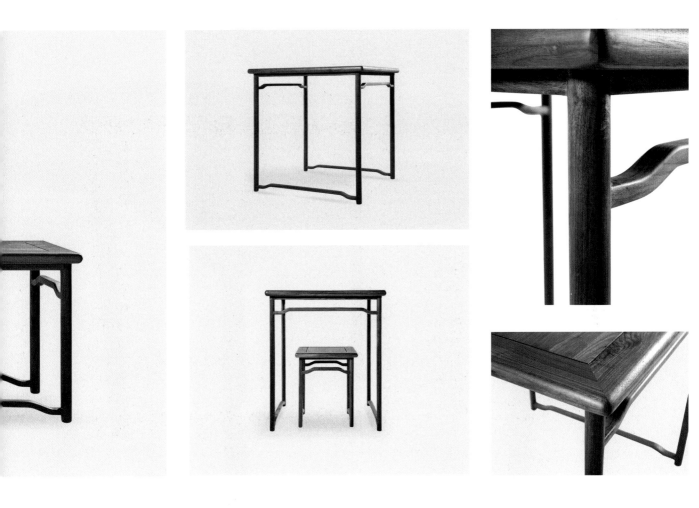

设计师在对中国传统家具的追溯与解读中感悟到了明式家具中所具有的"精进、文儒"的精神，将这种文化内涵融入传统家具再设计，在结构上追求极致简约，以"高低枨"的设计打破传统方桌四面完全对称的形式，力求放大、突出空灵的视觉效果。该方桌用传统手工制作技艺完成。

The craftsman comprehends the "enhancement and elegance" from the Ming-style furniture from the tracing and understanding of traditional Chinese furniture, and integrates such cultural connotation into redesign of traditional furniture. The craftsman pursues extreme simplicity, and breaks the full symmetry of traditional square table on four sides by the design of "high and low jambs" to magnify and highlight the vacant visual effect. The square table is made by traditional handcraftsmanship.

民以食为天 Food is God | 104

《晒桌》/ 综合材料八仙桌
作者：赵宇　糜思尧
材质：金属、竹、漆

Drying Table / Comprehensive Materials Square Table
Designers: Zhao Yu, Mi Siyao
Materials: Metal, Bamboo, Lacquer

中国饮食器物设计作品集　Design of Chinese Dining Utensils | 105

　　设计灵感来源于中国农村在丰收时节晒粮食这一习俗，设计元素采用了八仙桌（金属材质）和簸箕（竹编）。用现代设计中的解构手法来阐释传统的八仙桌，用中国漆艺的手法来修饰簸箕，体现了传统与现代的结合与包容。

The design idea is from the convention of drying grains in the sun in harvest seasons in rural area of China. The design adopts elements of square table (metal) and dustpan (bamboo). The deconstruction method in modern design is used to interpret the traditional square table. The dustpan is embellished by Chinese lacquer art, showing the combination and containment of tradition and modern.

《草木缘》/ 修复八仙桌
作者：敖进 杨承颖 刘潇
材质：木、稻草、糯米、坚果壳等综合材料

Affection of Straw and Wood / Repaired Square Table
Designers: Ao Jin, Yang Chengyin, Liu Xiao
Materials: Wood, Straw, Sticky rice, Nuts, Comprehensive materials

八仙桌是用于吃饭饮酒的家具,可以围坐八个人,故百姓俗称"八仙桌"。

此作品原型为一传统的八仙桌,由于岁月长久其已腐烂。作者利用其腐旧形成的肌理,使用具有黏稠性的糯米和稻草、坚果等材料结合对木桌进行修复。通过打磨让稻草若隐若现,使木桌在与"草"的结合中得以重生,是为"草木结缘"。

Square table is a piece of furniture used for dining and drinking. Eight people can sit around the table, so it is commonly known as "eight-immortal" table.

The original shape of the work is a traditional square table, which has been decayed after a very long time. The craftsmen make use of the texture formed by decomposition to repair the wood table by sticky rice, straws and nuts. By polishing, the straws are indistinct. The wood table revives via combination with the straws. So it is named "affection of straw and wood".

"竹八仙"(Bamboo Square Table)用碳化后的楠竹,利用竹材自身的物理特性,以扭曲、热弯成型,通过中国传统的榫卯连接方式完成。

为稳定桌腿和桌面的结构,以横杖与曲杖穿插,不仅加强了桌的稳定性,并以腿为中心构成中国青铜器上的饕餮纹简化图案。桌面用透雕的手法镂空出矩阵图案,丰富了视觉也呼应了其他部位副空间。

The "Bamboo Square Table" is made by carbonized phyllostachys pubescens. It adopts traditional Chinese tenon-and-mortise connection and modeling of distortion and hot bending by making use of the physical characteristics of bamboos.

In order to stabilize the structure of legs and table top, the horizontal rod and bended rods are interspersed to not only strengthen the stability of the table but also constitute the simplified Taotie pattern on bronze ware. The table top adopts deep carved method to hollow out matrix pattern, and enrich the vision and also echoes to the subspace of other area.

《竹八仙》/ 全竹八仙桌
作者: 张海涛
材质: 竹

Bamboo Square Table / Full-bamboo Square Table
Designer: Zhang Haitao
Material: Bamboo

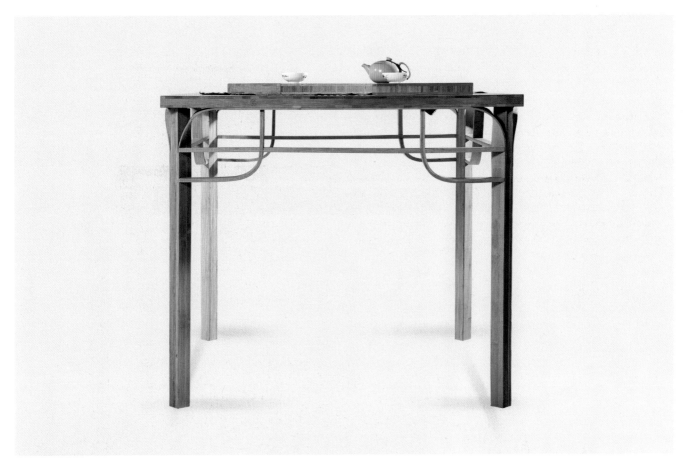
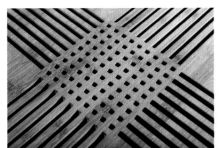

民以食为天 Food is God | 110

《新中式实木八仙桌》之一
作者：谢垚
材质：木

Neo-Chinese Style Solid Wood Square Table I
Designer: Xie Yao
Material: Wood

　　明代家具中的"剑腿"为本设计带来了启发,设计剔除掉传统家具中常见的烦琐雕刻元素,取其清秀挺拔之态,用其智慧巧妙的结构,以最简练的手法制作完成,使之能够在现代家居环境中和谐配搭。

The "sword leg" in furniture of Ming Dynasty enlightens this design. The design removes the common tedious sculpture elements in traditional furniture and takes the state of delicate and straight. This table is finished by smart and artful structure and the simplest technique so that the table can be harmonious in modern home environment.

这张八仙桌灵感来自明代的棋牌桌。该设计将牙板和腿结合在一起,将装饰和结构结合在一起。同时,薄薄的抽屉显得精致灵动。

The design idea of this square table is from the chess table in Ming Dynasty. The design combines dental plate and legs, and combines decoration and structure. The thin drawer is exquisite and flexible.

《新中式实木八仙桌》之二
作者:谢垚
材质:木

Neo-Chinese Style Solid Wood Square Table II
Designer: Xie Yao
Material: Wood

天然石材经现代工艺加工处理后能够散发出独特的视觉魅力。"交互石代"使用颜色和花纹都对比强烈的两种石材通过几何图形的相互切割、拼贴的方式制作，营造出"你中有我，我中有你"的交互视觉效果。结构上采用易拆卸的连接方式组合，造型简洁、时尚，极具现代感。

The nature stone, after processing by modern technology, can express unique charm of vision. The "Interactive Stone Age" is made by two kinds of stones with strong contrast of colors and patterns by means of cutting and tiling of geometric figures to create the interactive visual effect of "you have me and I have you". The "Interactive Stone Age" adopts a structure of connection easy to dismantle. The model is simple, fashion and modernized.

《交互石代》
作者：吴菡晗　田棱锐
材质：大理石

Interactive Stone Age
Designer: Wu Hanhan, Tian Lingrui
Material: Marble

《高山流水》/ 实木多用途八仙桌
作者：赵卫东　彭科星　罗昊　马林明
材质：木

Mountain & Stream / Solid Wood Multipurpose Square Table
Designers: Zhao Weidong, Peng Kexing, Luo Hao, Ma Linming
Material: Wood

"品味中国·牵手世界"展览以中国"坝坝宴"的形式呈现。而"坝坝宴"最典型的饮食方式是"流水式的高潮"不间断重复。"高山流水"八仙桌设计概念,就是来自"高潮起伏,源源不断"的含义。

　　传统"书卷"与层叠起伏的"山体"结合的托盘造型象征着"高山";飞流直下的桌旗造型象征着"流水","高山流水"八仙桌,隐喻出以"流水式高潮重复"为饮食方式的"坝坝宴"的展览形式主题。

The exhibition of "Taste China · Connects with the World" is presented by form of "Nine Steamed Dishes" in China. While the most typical dietary mode of "Nine Steamed Dishes" is "running-water climax" and intermittent repetition. The design concept of "Mountain & Stream" square desk is from the meaning of "running-water climax and continuousness".

The shape of tray combining traditional "scrolls" and cascading mountains stands for "high mountains"; the zigzagging table flag like waterfalls stands for "streams". The "Mountains & Stream" square table shows a metaphor of dietary mode of "running-water climax" and intermittent repetition, and the theme of exhibition way of "Nine Steamed Dishes".

《糙木茶桌》
作者：王立端 李其龙 糜思尧
材质：原木，大漆

Rough Wood Tea Table
Designers: Wang Liduan, Li Qilong, Mi Siyao
Materials: Wood, Lacquer

中国饮食器物设计作品集 Design of Chinese Dining Utensils | 117

　　中国自古都有用原木制作家具的传统，这款糙木茶桌的造型具有功能交融于自然、粗犷对比出精细的特点。传统手工打磨的漆艺与天然糙木的搭配增强了材质的对比，也使该茶桌具有了唯一性。

China has a long history to make furniture by logs. The model of this tea desk made by rough wood integrates functions into nature, and shows exquisiteness contrasting with roughness. The lacquer art polishing by traditional workmanship matches the nature rough wood to enhance the contrast of materials, and make the tea table unique.

《舞》/ 综合材料八仙桌
作者:张海东
材质:木、毛线、金属

Dance / Comprehensive Materials Square Table
Designer: Zhang Haidong
Materials: Wood, Kitting wool, Metal

中国饮食器物设计作品集 Design of Chinese Dining Utensils | 119

该作品以中国传统饮食文化及民俗文化为灵感，将老式八仙桌与现代纤维艺术相结合。纤维以红色为主，凸显"中国红"元素，穿插其中的纤维材料主要混合了天然羊毛线及各种人造纤维，并由金属丝等硬质材料固定造型。纤维块面以"S"形将桌子一分为二，也暗合了太极八卦的图形。材质与造型的多重选择，体现了文化由传统向现代过渡的形态。

The design idea of this work is from traditional Chinese food culture and folk culture. It combines the old-fashioned square table with modern fiber art. The fibers are mainly in red, showing the element of "China Red". The interspersing fiber materials contain nature wool and various artificial fibers, and fixed by hard materials such as metal wire. The fiber surface separates the table into two parts by an "S" shape line, which also implies the figure of eight-diagram-shaped appetizer. The multiple choices of materials and models show the form of culture transiting from tradition to modern.

民以食为天 Food is God | 120

　　作品以竹制家具为基础，与中国西南少数民族地区水族的传统刺绣工艺结合，尝试用水族的传统几何吉祥纹样在竹片上作绣。刺绣以水族的蓝色为主色调，红、黄色为辅，竹子的硬度和棉线的柔软形成鲜明对比，呈现出绚丽多彩的民族风味和视觉效果。

The work is based on the bamboo furniture, and is combined with the traditional embroidery of the Shui Nationality in the ethnic minority area of Southwest China. The traditional geometric auspicious patterns of the Shui Nationality are embroidered on bamboo chips. The embroidery features the blue typical of the Shui Nationality. Red and yellow are adopted as the supplementary colors. The hardness of bamboos and the softness of cotton threads form a striking contrast, and create a bright and colorful nationality taste and visual effect.

《族》
作者：潘宏甲
材质：竹、毛线

Nationality
Designer: Pan Hongjia
Materials: Bamboos, Woolen yarns

　　静直案几灵感来源于中国传统建筑的斗拱结构。斗拱之间有规律的交错形成了线的韵律节奏，不仅是形式上的美感，也是结构上的需要。简洁的无曲线的形式让静直案几在现代主义和中国传统中取得了一个平衡点。

The inspiration comes from the system of brackets inserted between the top of a column and a crossbeam in the traditional Chinese buildings. The regular intersections between the system of brackets inserted between the top of a column and a crossbeam form linear rhythms. This is for the sake of not only form aesthetics, but also structural needs. The simplified non-curvilinear form seeks a balance for the quiet and straight cases between the modern creative ideas and the Chinese traditions.

《静直案几》
作者：谢垚　张庆
材质：实木

Quiet and Straight Case
Designers: Xie Yao, Zhang Qing
Material: Solid wood

顾名思义,该系列主要致力于对边缘的各种研究和尝试。该餐桌边缘弧度经过精心把控,桌面与桌腿之间也以圆弧过渡,桌面有食盘设计,食盘与桌面实为整体打磨而成,其边缘极细,食盘以外桌面打磨圆润光滑,食盘内部手工凿挖肌理,形成触觉和视觉上的对比,与月球表面肌理有一定相似之处。

Just as its name implies, the series is based on research of the edge. The edge of the dining table is exquisitely controlled. The desk surface and the desk leg are also connected by arcs. On the surface of the dining table, the dining tray is designed. The food tray and the table surface are polished and integrated into a whole. Its edge is fine. Outside the service plate, the desk surface is polished to be extremely smooth. The inside of the service plate is chiseled by hand to form a contrast of touch and vision. The texture of the inside surface resembles the surface texture of the moon.

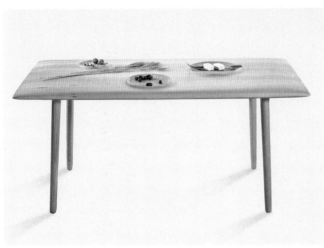

《沿》
作者:李其龙
材质:实木

Edge
Designer: Li Qilong
Material: Solid wood

此展架设计灵感来源于中国传统条凳造型,结构上采取模块化结构设计,将展板支架分割成类似梯子一样的模块,高低搭配可以随意变换设计造型,且方便运输。设计凸显出清新、自然的感觉。

The design idea of the display rack is from the model of traditional Chinese bench. The design adopts modularized structure to separate the rack support into modules like ladders. The high and low rack can be matched and the model can be changed arbitrarily. Besides, it is convenient for transportation. The design shows the feeling of freshness and nature.

《组合展架》
作者:曹悦
材质:木材

Combination Display Rack
Designer: Cao Yue
Material: Wood

作品主题定位于大自然,材料取自自然界随处可见的油菜花。将油菜籽、菜花、菜壳和菜杆分别放入高低不同的树脂模具中,希望将这些易逝的美丽事物凝于作品中,另一方面也打破了油菜花、油菜籽的传统表现形式。在设计理念上,以油菜花的生长到结束的过程形态为基点,意味着虽然生命是不断变化发展到最后消逝,但仍能留住其中的美好回忆时光。

The work is positioned to be natural. The materials contain rape flowers which are everywhere to find in the nature. The rapeseeds, rape flowers, shells and stems are inserted into resin molds of different heights. It is hoped that these transient beautiful things are all fixed on the work. On the other hand, the traditional expression form of rape flowers and rapeseeds is innovated. In terms of design concepts, the process from the birth to the death of rape flowers is adopted to imply that life keeps growing and changing to the final moment, and the beautiful memories will always remain.

《芸薹未了》
作者:汪梅 冯唐菡 曾贤思 周璐丹 王鸿
材质:树脂

Gone Winter Rapes
Designers: Wang Mei, Feng Tanghan, Zeng Xiansi, Zhou Ludan, Wang Hong
Material: Resin

文化传承的手工艺教学创新实践

| 韦 芳

在漫长的历史进程中,一个民族、一个国家的文化基因,往往一方面以建筑、器物等物质形态得以延续、传承;另一方面以人们的技艺、行为习惯和社会的风俗等非物形态得以发展、变迁。在整个社会进入现代工业标准化规模生产以前,人们所掌握的生产与生活技艺(手工艺)在习惯与风俗影响下,不断创造且丰富着人们需要的物质文明,也承载了当时人们文化取向与审美趣味聚集而成的精神文明,通过这种方式实现文化传承并推动着社会进步。可以说,手工艺者在相当长一段历史时期内,是民族文化的主要传承者、创新者、生产者之一。手工技艺通过一代一代的师徒传授得到传承、发展。

鸦片战争以后,中国进入长达近两百年的衰落期,西方文明在枪炮的裹挟下涌入中国。强势文明与弱势文明的对抗,既是强势文明对弱势文明政治、军事、经济、社会的全面冲击,更是对整个国家文化自信的深重打击。而随后被迫开始的西方标准的现代化进程,摧毁式地消解了手工艺行业的物质生产与文化传承的主力军角色,手工技艺传承也因为师徒传授教育模式的式微与西方现代教育体系在中国的确立而被大大削弱。直到今天,我们"学"必称西方依然是主流;直到现在,我们一个个手工技艺成为了"非遗"。

聚居37个少数民族的文化丰富性,高原、山地与平原兼具的地理丰富性,高海拔山地、温热带盆地、亚热带雨林皆有的气候丰富性,形成了中国西部腹地特殊的区位文化优势,共同滋养造就了身处其间的四川美术学院(川美)植根本土的文化自觉,致力传承创新的实践自觉。川美70多年的发展历程,以染织、漆艺等手工艺专业的历史最为悠久,积累了丰富的手工艺传承在高等院校的教学实践经验。今天,物质文明繁荣到一定程度,文化复兴与文化自信重建要求愈加迫切,市场多样化与小众化定制需求愈加增大,手工艺迎来了最好的时代。然而成为"非遗"本身,就是中国手工艺发展现状不适应时代的表征。手工技艺的传承延续性不良,手工技艺的现代应用性不良,手工艺从业者的创造与生产适应性不良,手工艺从业者的生存状况不良,形成了整个手工艺行业的发展不良。川美一众师生以手工艺的复兴为己任,从教学实践创新的角度切入,构建手工艺创新设计人才培养体系,开展多元模块与多种资源协同创新的教学实践活动,不断累积成果精品与优秀创新设计人才。我们期待通过优秀手工艺创新设计人才输出、成果精品展览、成果市场转化等方式,推动手工技艺传承创新,促进手工艺行业发展,助力中国的文化软实力提升与国民文化自信心提升。

川美的手工艺创新设计人才培养体系,主要是由基本教学模块、国际国内名匠工作坊、大师进校园协同创新项目、手工艺"双创平台"及国际化高端推广平台等部分构建而成的五位一体校内校外联结式系统化教学模式。

一、以专业工作室建构基本教学模块

从2008年开始,川美启动了手工艺的专业教学改革,建立了以漆艺、陶艺、纤维、金属为核心的专业工作室制人才培养模式实践。这种模式由基础通识教学、专业通识教学和专业工作室教学三个阶段共同构成,基础通识教学与专业通识教学阶段实行全员必修,专业工作室采取一定标准下的师生"双向选择"方式分配学生进入不同的工作室进行持续性和系统化的学习、创作，实现手工技艺的深入研习。

二、以国际国内名匠工作坊推动顶尖手工技艺的推广与传承

设立名匠工作坊，每年邀请各类国际国内顶尖手工艺大师到校开设不少于两周的工作坊课程，每次配以选拔机制面向师生征集工作坊学员，共同就特定的手工技艺传承开展研讨式学习，强调基于传统文化浸润与国际视野拓展的创新设计过程体验，推动顶尖手工技艺的推广与传承。

三、以"大师进校园"协同创新项目推动民间手工艺大师与院校师生的协同创新

开展"大师进校园"项目，与西南地区各省市工艺美术行业协会合作，选拔优秀民间工艺大师到川美，与师生结成研创团队，历时半年，通过集中授课、团队研讨、工坊创作三个阶段，共同完成作品创作。这一项目成果突出，受到了各方的肯定，并因此获得了2015年的国家艺术基金人才培养项目资助，探索了一条民间工艺大师的手工技艺创新发展路径。

四、以"双创平台"推动作品的产品属性转化与优秀人才的大师属性升级

依托川美微型企业园植入"教学带动创新，创新培育创业，创业助力产业升级"的创新创业教育内容。一方面，定期选拔有兴趣爱好的学生，组织创新创业训练营、科技前沿训练营等，挖掘手工艺创新性传承人才，培育、孵化具备市场前景的作品、创业项目；另一方面，引入风投推动作品的产品属性转化与优秀人才的大师属性升级，实现创新成果的市场转化。

五、以重大国际推广项目推动手工艺创新精品为载体的中国文化国际交流

通过基础教学模块、国际国内名匠工作坊、"大师进校园"协同创新项目以及"双创平台"等四个方面的长期教学实践，累积了一批承载中国文化的手工艺创新设计精品。以促进中国文化的国际交流与沟通为目的，以手工艺创新设计精品为载体，充分运用米兰世博会、中国国家艺术基金海外传播项目等高端推广渠道，开展中国文化的国际交流活动，助力国家文化软实力水平提升。

Cultural Heritage Through the Innovation Practice in Handicraft Education

| Wei Fang

In the years of evolution, the cultural genes of a people or a nation are on one hand inherited from one generation to another through the tangible formation, including architectures and artifacts, and on the other hand evolve through intangible formation, namely crafts, behavior habits, and customs. Before industrialization and standardization of manufacture, the production and traditional technology (handicraft), in the effect of tradition and customs, are the ways to create and fertilize the tangible civilization, providing comfort for people's life, carrying on the intangible civilization formed by people's cultural orientation and aestheticism, and consequently realizing cultural inheritance and facilitating social development. In such a perspective, craftsmen have been in history the major heirs, innovators and manufacturers of national culture, thanks for whom the handicraft can be carried on by way of apprenticeship from generation to generation.

Since the Opium War, China had been falling for nearly two centuries in the invasions of the western civilization. The confrontation between the strong civilization and the weak one is not only an overwhelming jolt to politics, military, economy, and society but a disastrous punch on the self-esteem of national culture. In the wake of the confrontation, the forced modernization with the western standards has crushed the handicraft industry as a pillar of tangible production and intangible heritage, the heritage of craftsmanship fading away between the fall of apprenticeship and the rise of modern education system. As a result, we consecrate western culture and ignore our own handicrafts, which are bound to fade away unless they are entitled "intangible heritage".

The diversity in cultures, landforms and climate shapes the special regional culture of inland of west China, where inhabit 37 ethnic peoples, exhibit landforms of plateaus, mountains, and plains, and present various climates of high mountain, temperate basin, and sub-tropical rain forest. Such a lavishness of culture and nature nourishes Sichuan Fine Arts Institute located here of its cultural and practical awareness of localization and its commitment to innovation and heritage. In its history of over seven decades, SCFAI, especially the two majors of Dyeing and Weaving Art and Lacquer Art, has accumulated rich experience in cultural heritage though handicraft education. In the times of high prosperity of material civilization, with a pressing need of cultural renaissance and cultural self-esteem restoration, we usher in the best times of handicraft, the market diversified and the need mounting in customized production. Nevertheless, being "intangible heritage" itself indicates that Chinese handicraft is lagged behind the times, the handicraft unable to be inherited and the craftsmanship unable to provide artisans a decent life, both of which result in the underdevelopment of handicraft industry. Taking on themselves the duty of handicraft rejuvenation, the teachers and students of SCFAI started with innovation of design education, building cultivation system of innovation talents, implementing teaching activities integrating multi-modes with multi-resources

for innovation, making achievements in quality designs and creative designers. We expect to power the handicraft heritage and innovation, to facilitate the development of the industry, and to assist the upgrade of the soft power of Chinese culture and the self-esteem of Chinese people, through the output of creative designers, the design exhibitions, and the marketization of artistic works.

The cultivation system of creative designers is integrate of resources on-campus and off-campus, or a 5-in-1 teaching mode, including the basic teaching modules, the workshops of prominent artisans international and national, the "Masters coming to campus" projects of collaborative innovation, "the innovation-entrepreneurship platform" for handicraft, and international high-end promotion platform.

Ⅰ. Basic Teaching Modules based on professional studios

Since 2008, SCFAI has launched the reform of specialty teaching, starting

Trial in cultivation module of studio designers mainly of lacquer art, pottery art, fiber, and metal. The module is composed of three phases, liberal education, professional education, and professional studio education. The liberal education and professional education provide compulsory courses, while professional studio offers a "two-way choice" between teachers and students under certain guidelines, students assigned to different studios for further study and creation in hope of in-depth study of the handicraft.

Ⅱ. Facilitation for promotion and heritage of top crafts through the workshops of prominent artisans international and national

SCFAI set up artisan workshop, inviting top artisans both international and national to hold workshop no less than two weeks per year. During the workshop, a selection scheme works to recruit trainees among teachers and students, who will participate in seminars of the heritage of a certain handicraft, highlighting the process experience of creative design based on both traditional culture and global horizon, the scheme facilitating the promotion and heritage of top handicraft.

Ⅲ. Facilitation for collaborative innovation between folk artisans and faculty-student through the project of "Masters Coming to Campus"

SCFAI launched the project of "Masters Coming to Campus", collaborating with the societies of crafts and fine arts in Southwest provinces, selecting prominent folk craft masters to team up with faculties and students for R&D, co-working on designs in six months through three phases of lecturing, discussion, and workshop creation. The project is fruitful, winning praises in concerned parties and, more important, enjoying the funding from National Art Fund in 2015 for the project pioneering a new path of folk craft masters' innovation.

Ⅳ. Promotion based on "Innovation-Entrepreneurship Platform" of the marketization of artistic works and the upgrade from designers to masters

Relying on SCFAI miniature innovation park, the education is aimed at innovation and entrepreneurship with the guideline of "teaching driving innovation, innovation breeding entrepreneurship, entrepreneurship facilitating the upgrade of the industry." On one hand, the school periodically selects students for the training camp of innovation and entrepreneurship or the training camp of sci-tech-front so as to discover creative designers, nurse and incubate artistic works and entrepreneurship projects with a promising market. On the other hand, the school brings in venture capital, driving the marketization of works and the upgrade from designers to masters.

Ⅴ. Facilitating international exchanges of Chinese culture with the quality handicraft works as a carrier through major global promotion project

The school has accumulated a lot of quality designs of handicraft through the teaching practice in four aspects of the basic teaching modules, the workshops of prominent artisans international and national, the project of "Masters Coming to Campus", and "the Innovation-Entrepreneurship Platform". We are committed to enhance the culture soft power of China, aiming at more international exchanges of Chinese culture, relying on quality handicraft designs, utilizing the high-end channels, like the Milan Expo and the project of oversea publicity by China National Art Fund, to facilitate international exchanges of Chinese culture.

编后记

2015年中外文化交流中心与四川美术学院和雅伦格文化艺术基金会(威尼斯)联合在意大利举办了两场中国饮食器物设计文化展,一场名为"品味中国·牵手世界",另一场名为"民以食为天"。展览大获成功,被列为2015年米兰世博会官方展览项目。中外110余家中外媒体予以了采访报道,展览为传播中国文化起到了积极作用。为在此基础上进一步扩大交流传播效果,2016年三家主办单位在前期展览基础上联合申报国家艺术基金海外传播推广资助项目,成功获批立项。在巴黎中国文化中心与布鲁塞尔中国文化中心的支持和参与下,这批设计作品将以《"丝路长·宴四方"——中国饮食器物设计文化展》之名于2017年4月-6月在法国巴黎和比利时布鲁塞尔展出。之后还将赴土耳其巡展。本作品集所刊载的是作者们近几年有深入地域传统手工艺的留存地、在地调研、在地实践、在地创意制作完成的作品。希望这些作品能将其所特有的乡土之味、材料之朴、工艺之巧,传递给欧洲观众。同时希望通过对造物精神的弘扬,让民族的、传统的、地域的文化可持续发展。

Afterword

Center of International Cultural Exchange, Sichuan Fine Arts Institute and EMGdotART Foundation jointly hosted twice "Chinese Tableware Design and Culture Exhibitions "in Italy in 2015, separately named as "Taste China•Connect the world" and "Food is God". The exhibitions were successful and listed as two of the official exhibition items in Milan Expo 2015. There were over 110 domestic and foreign medias reported the event and both exhibitions played an extremely positive role in sharing Chinese culture across the world. In order to expand the media effect to a higher level, in 2016 the three organizers jointly applied for the overseas-spreading funding project sponsored by the China National Arts Fund and was approved. In the name of "Feast along the Silk Road—An Exhibition of Chinese Tableware Design and Culture" the works will be exhibited in Paris, France and Brussels, Belgium during April to June 2017, and in Turkey soon afterwards. In this issue, the great designs are the collections of the artists who researched and learned among those locations where almost lost the inheritance of the productions through their hard works. Every piece of design not only showed the original designers' creativities and it also shown their talent as well. We hope our European audiences will appreciated these art works through the culture taste, the pure down to the earth materials and the detail of the productions. In addition we hope we will continually sharing and promoting of The Spirit of Creation to provide historical tradition, folk crafting, and cultural inheritance to the world.

附录

一、米兰站 展览名：《"品味中国·牵手世界"——中国饮食器物设计文化展》
　主办单位：中外文化交流中心　四川美术学院
　　策展人：段胜峰　汤静　邓良军　　学术主持/设计主持：王立端

二、威尼斯站 展览名：《民以食为天——中国饮食器物设计文化展》
　主办单位：中外文化交流中心　四川美术学院　雅伦格文化艺术基金会
　　策展人：马里诺·福林　王立端　　联合策展人：段胜峰　汤静　李家豪　邓良军　吴菡晗

三、巴黎站 展览名：《"丝路长·宴四方"——中国饮食器物设计文化展》
　主办单位：巴黎中国文化中心　中外文化交流中心　四川美术学院　雅伦格文化艺术基金会
　　策展人：王立端　　　　联合策展人：段胜峰　汤静　李家豪　邓良军
　　政府支持：中国驻法国大使馆文化处　重庆市文化委员会
　　赞助：ART-ZF 樊尚创意企业发展有限公司　重庆宏美达欣兴实业（集团）有限公司

四、布鲁塞尔站 展览名：《"丝路长·宴四方"——中国饮食器物设计文化展》
　主办单位：布鲁塞尔中国文化中心　中外文化交流中心　四川美术学院　雅伦格文化艺术基金会
　　策展人：王立端　　　　联合策展人：段胜峰　汤静　李家豪　邓良军
　　政府支持：中国驻比利时大使馆文化处　重庆市文化委员会
　　赞助：ART-ZF 樊尚创意企业发展有限公司　重庆宏美达欣兴实业（集团）有限公司

Appendix

Ⅰ. Milan Station "Taste China·Connect the World"–Design Exhibition of Chinese Dining Utensils
Co-presented by Center of International Cultural Exchange　Sichuan Fine Arts Institute
Curators: Duan Shengfeng, Tang Jing, Vincent L.J. Deng　　　Academic/Design director: Wang Liduan

Ⅱ. Venice Station "Food is God–Design Exhibition of Chinese Dining Utensils"
Co-presented by Center of International Cultural Exchange　Sichuan Fine Arts Institute　EMGdotART Foundation
Chief Curators: Marino Folin, Wang Liduan
Curators: Duan Shengfeng, Tang Jing, Victor Li Vincent, L.J. Deng, Wu Hanhan

Ⅲ. Paris Station "Feast along the Silk Road–An Exhibition of Chinese Tableware Design and Culture"
Co-presented by Centre Culturel de Chine à Paris Center of International Cultural Exchange
Sichuan Fine Arts Institute　EMGdotART Foundation
Chief Curator: Wang Liduan　　Curators: Duan Shengfeng, Tang Jing, Victor Li Vincent, L.J. Deng
Patrons: Cultural Office of the Embassy of the People's Republic of China in France
Chongqing Municipal Culture Commission
Sponsors: ART-ZF　VINCE CREATIVE DEVELOPMENT CO,.LTD.　Chongqing Wintus(New Star)Enterprises Group

Ⅳ. Brussels Station "Feast along the Silk Road–An Exhibition of Chinese Tableware Design and Culture"
Co-presented by China Cultural Center in Brussels Center of International Cultural Exchange
Sichuan Fine Arts Institute　EMGdotART Foundation
Chief Curator: Wang Liduan　　Curators: Duan Shengfeng, Tang Jing, Victor Li Vincent, L.J. Deng
Patrons: Cultural Office of the Embassy of the People's Republic of China in the kingdom of Belgium
Chongqing Municipal Culture Commission
Sponsors: ART-ZF　VINCE CREATIVE DEVELOPMENT CO,.LTD.　Chongqing Wintus(New Star)Enterprises Group

图书在版编目（CIP）数据

民以食为天：中国饮食器物设计作品集/王立端，段胜峰主编. —重庆：重庆大学出版社，2017.3
ISBN 978-7-5689-0444-5

Ⅰ.①民… Ⅱ.①王… ②段… Ⅲ.①餐具—设计—作品集—中国—现代 Ⅳ.①TS972.23

中国版本图书馆CIP数据核字（2017）第043842号

民以食为天：中国饮食器物设计作品集

MIN YI SHI WEI TIAN：
ZHONGGUO YINSHI QIWU SHEJI ZUOPINJI

主　编　王立端　段胜峰

副主编　汤　静　韦　芳　谢亚平

策划编辑：张菱芷

责任编辑：张菱芷　　　版式设计：汪　泳　王佳琪
责任校对：邹　忌　　　责任印制：赵　晟

重庆大学出版社出版发行
出版人：易树平
社址：重庆市沙坪坝区大学城西路21号
邮编：401331
电话：（023）88617190　88617185（中小学）
传真：（023）88617186　88617166
网址：http://www.cqup.com.cn
邮箱：fxk@cqup.com.cn（营销中心）
全国新华书店经销
重庆新金雅迪艺术印刷有限公司印刷

开本：889mm×1194mm　1/16　印张：8.75　字数：479千
2017年4月第1版　2017年4月第1次印刷
ISBN 978-7-5689-0444-5　定价：168.00元

本书如有印刷、装订等质量问题，本社负责调换
版权所有，请勿擅自翻印和用本书
制作各类出版物及配套用书，违者必究